P9-CJT-059

MOLDING OF PLASTICS

MOLDING OF PLASTICS

Encyclopedia Reprints

Edited by

NORBERT M. BIKALES

Consulting Chemist
Livingston, New Jersey

WILEY-INTERSCIENCE

a Division of John Wiley & Sons, Inc.
New York • London • Sydney • Toronto

GLOSSARY, reprinted from *Encyclopedia of Polymer Science and Technology*, Vol. 9, pp. 1–9

COMPRESSION AND TRANSFER MOLDING, reprinted from *Encyclopedia of Polymer Science and Technology*, Vol. 9, pp. 9–47

INJECTION MOLDING, reprinted from *Encyclopedia of Polymer Science and Technology*, Vol. 9, pp. 47–83

BLOW MOLDING, reprinted from *Encyclopedia of Polymer Science and Technology*, Vol. 9, pp. 84–118

ROTATIONAL MOLDING, reprinted from *Encyclopedia of Polymer Science and Technology*, Vol. 9, pp. 118–137

AUXILIARY PROCEDURES, reprinted from *Encyclopedia of Polymer Science and Technology*, Vol. 9, pp. 137–157

BAG MOLDING, reprinted from *Encyclopedia of Polymer Science and Technology*, Vol. 2, pp. 300–316

MOLDS, reprinted from *Encyclopedia of Polymer Science and Technology*, Vol. 9, pp. 158–181

RELEASE AGENTS, reprinted from *Encyclopedia of Polymer Science and Technology*, Vol. 12, pp. 57–65

Library of Congress Catalog Card Number: 78-172950

ISBN 0-471-07233-8

Printed in the United States of America.

10 9 8 7 6 5 4 3 2 1

CONTRIBUTORS

NORBERT M. BIKALES, Consulting Chemist, Livingston, New Jersey, *Editor*

RICHARD E. DUNCAN, U.S. Industrial Chemicals Company, Tuscola, Illinois, *Rotational Molding*

DAVID R. ELLIS, U.S. Industrial Chemicals Company, Tuscola, Illinois, *Rotational Molding*

A. B. HITCHCOCK, Western Electric Company, Indianapolis, Indiana, *Auxiliary Procedures*

JOHN L. HULL, Hull Corporation, Hatboro, Pennsylvania, *Glossary; Compression and Transfer Molding*

GEORGE P. KOVACH, Koro Corporation, Hudson, Massachusetts, *Release Agents*

ROBERT A. MCCORD, U.S. Industrial Chemicals Company, Tuscola, Illinois, *Rotational Molding*

SAM MONROE, General Dynamics Corporation, Fort Worth, Texas, *Bag Molding*

G. E. PICKERING, Arthur D. Little, Inc., Cambridge, Massachusetts, *Blow Molding*

SUMNER E. TINKHAM, Tech-Art Plastics Company, Morristown, New Jersey, *Molds*

LEE J. ZUKOR, Allied Testing and Research, Hillsdale, New Jersey, *Injection Molding*

PREFACE

Molding is one of the critical steps in converting polymers to a myriad of useful objects. An idea of the importance of the subject can be gained from the fact that more than 30,000 compression- and transfer-molding machines, 40,000 injection-molding machines, and 2,000 blow-molding machines are in operation in the United States.

This book assembles in one place nine chapters from three different sources dealing with the various molding methods, the auxiliary procedures related to molding, and mold design. The contributions are by recognized authorities in the field and this book, together with that entitled *Extrusion and Other Plastics Operations*, issued at the same time, should provide a comprehensive description of modern processing of plastics.

Because the chapters are reproduced unchanged from their original sources, the reader should ignore the cross references to other pages and to other chapters that appear in the text.

I should like to acknowledge the great help of Jo Conrad in editing the original Encyclopedia articles on which this book is based.

NORBERT M. BIKALES

Livingston, New Jersey
June 1971

CONTENTS

GLOSSARY 1

Reprinted from *Encyclopedia of Polymer Science and Technology*, Vol. 9, pp. 1–9

COMPRESSION AND TRANSFER MOLDING 11

Reprinted from *Encyclopedia of Polymer Science and Technology*, Vol. 9, pp. 9–47

INJECTION MOLDING 51

Reprinted from *Encyclopedia of Polymer Science and Technology*, Vol. 9, pp. 47–83

BLOW MOLDING 89

Reprinted from *Encyclopedia of Polymer Science and Technology*, Vol. 9, pp. 84–118

ROTATIONAL MOLDING 125

Reprinted from *Encyclopedia of Polymer Science and Technology*, Vol. 9, pp. 118–137

AUXILIARY PROCEDURES 145

Reprinted from *Encyclopedia of Polymer Science and Technology*, Vol. 9, pp. 137–157

BAG MOLDING 167

Reprinted from *Encyclopedia of Polymer Science and Technology*, Vol. 2, pp. 300–316

MOLDS 183

Reprinted from *Encyclopedia of Polymer Science and Technology*, Vol. 9, pp. 158–181

RELEASE AGENTS 207

Reprinted from *Encyclopedia of Polymer Science and Technology*, Vol. 12, pp. 57–65

INDEX 217

MOLDING OF PLASTICS

GLOSSARY

The terms in the glossary that follows were selected primarily from a comprehensive list of definitions of terms pertaining to plastics, "Glossary of Plastics Terms; A Consensus," *Plastec Note 14*, Plastics Technical Evaluation Center, Picatinny Arsenal, Dover, N.J., 1966. Other terms were taken from the article "Extrusion Die Nomenclature," *SPE Journal* **23** (8), 47 (1967); definitions of terms pertaining to blow molding were supplied by Mr. George Pickering, the author of the section on Blow Molding.

Accumulator. An auxiliary ram extruder which is used to provide extremely fast parison delivery. The accumulator cylinder is filled with plasticated melt coming from the extruder between parison deliveries and is stored or "accumulated" until the ram is required to deliver the next parison.

Autoclave molding. A modification of pressure bag molding in which after lay-up, the entire assembly is placed in steam autoclave at 50–100 psi. See BAG MOLDING.

Automatic mold. A mold for injection or compression molding that repeatedly goes through the entire cycle, including ejection, without human assistance.

Back draft. An area of interference in an otherwise smooth-drafted encasement; an obstruction in the taper which would interfere with the withdrawal of the model from the mold.

Back pressure. The resistance to the forward flow of molten material in an extruder.

Backing plate. A plate which backs up the cavity blocks, guide pins, bushings, etc, sometimes called support plate.

Bag molding (qv). A molding technique in which consolidation of the material in the mold is effected by the application of fluid pressure through a flexible membrane.

Blow pin. The pin (or needle) through which the blowing air is injected into the parison during blow molding.

Blow molding. A method of producing hollow objects by injecting a parison of hot melt into a hollow, two-part mold, then inflating the parison against the cool mold surface where it "freezes" into shape.

Blowup ratio. In blow molding, the ratio of the mold-cavity diameter to the parison diameter.

Breaker plate. A perforated plate located at the rear of the die in an extruder. It supports the screens that prevent foreign particles from entering the die. Also used without screens in injection-molding nozzles to improve distribution of color particles in the melt.

Breathing. The opening and closing of a mold to allow gases to escape early in the molding cycle; also called degassing.

Bulk factor. The ratio of the volume of a molding compound or powdered plastic to the volume of the solid piece produced therefrom; the ratio of the density of the solid plastic object to the apparent density of the loose molding powder.

Bushing. The outer ring of any type of a circular tubing or pipe die that forms the outer surface of the extruded tube or pipe.

Casting. The processes whereby a liquid polymer is poured into an open mold where it hardens without application of external pressure; also the finished product of a casting operation. The term is not to be used as a synonym for molding. See CASTING.

Cavity. The depression in a mold; the space inside a mold wherein a resin is poured; the female portion of a mold; that portion of the mold which encloses the molded article or which forms the outer surface of the molded article (often referred to as the die).

Chase. The main body of the mold which contains the molding cavity or cavities, or cores, the mold pins, the guide pins or the bushings, etc; an enclosure of any shape used to shrink-fit parts of a mold cavity into place, to prevent spreading or distortion in hobbing, or to enclose an assembly of two or more parts of a split-cavity block.

Choke. A plug or similar device located in a channel in a blow-molding manifold designed to divert and restrict flow to a desired path or to equalize pressures.

Clamping pressure. In injection molding and in transfer molding, the pressure which is applied to the mold to keep it closed; it acts in opposition to the fluid pressure of the compressed molding material.

Cold molding. The shaping of an unheated compound in a mold under pressure, followed by heating the article to cure it.

Compression mold. A mold which is open when the material is introduced and which shapes the material by heat and by the pressure of closing.

Compression ratio. In an extruder screw, the ratio of volume available in the first flight (at the hopper) to the last flight (at the end of the screw).

Contact molding. A process for molding reinforced plastics in which reinforcement and resin are placed on a mold and cured either at room temperature, using a catalyst–promoter system, or by heat in an oven; no additional pressure is used.

Core. A channel in a mold for circulation of heat-transfer media; part of a complex mold that forms undercut parts.

Core pin. A pin for forming a hole or opening in a molded piece.

Crosshead. A device which is attached to the discharge end of the extruder cylinder, designed to facilitate extruding material at an angle (normally, a 90° angle to the longitudinal axis of the screw).

Cull. The material remaining in a transfer chamber after the mold has been filled.

Cut-off. The line where the two halves of a compression mold come together; also called flash groove or pinch-off.

Deflashing. The operation in which the excess material of pinch-off and flash is removed from blown containers.

Degassing. See Breathing.

Die. An assembly of parts which, when attached to an extruder, is used to form an extrudate into a desired shape.

Double-ram press. A press for injection or transfer molding in which two distinct systems of the same kind (hydraulic or mechanical) or of a different kind, create, respectively, the injection or transfer force and the clamping force.

Double-shot molding. A means of producing two-color parts in thermoplastic materials by successive molding operations.

Dowel. A pin, fitted into one part of a mold, which enters a hole in the other part so that when the mold is closed the two parts become accurately aligned.

Draft. The taper or slope of the vertical surfaces of a mold designed to facilitate removal of molded parts.

Draft angle. The angle made by the tangent to the surface at that point and the direction of ejection.

Dwell. A pause in the application of pressure to a mold, made just before the mold is completely closed, to allow the escape of gas from the molding material.

Ejection. The process of removing a molding from the mold impression, by mechanical means, by hand, or by the use of compressed air.

Ejection plate. A metal plate used to operate ejector pins; designed to apply a uniform pressure to them in the process of ejection.

Ejection ram. A small hydraulic ram fitted to a press for the purpose of operating ejector pins.

Ejector. An attachment to an hydraulic press used for operating ejector pins. It may be mechanically, hydraulically, pneumatically, or electrically operated.

Ejector pin. A pin or thin plate that is driven into a mold cavity from the rear as the mold opens, forcing out the finished piece; also called knockout pin.

Ejector rod. A bar that actuates the ejector assembly when mold is opened.

Entrance angle. The maximum angle at which the molten material enters the land area of the die, measured from the center line of the mandrel.

Flash. The extra plastic attached to a molding; it must be removed before the part can be considered finished.

Flash gate. A long, shallow, rectangular gate.

Flash groove. A groove ground in a mold to allow escape of excess material.

Flash line. A raised line appearing on the surface of a molding and formed at the junction of mold parts.

Flash mold. A mold designed to permit the escape of excess molding material. Such a mold relies upon back pressure to seal the mold and put the piece under pressure. See MOLDS.

Flash ridge. That part of a flash mold along which the excess material escapes until the mold is closed.

Floating chase. A mold member, free to move vertically, which fits over a lower plug or cavity, and into which an upper plug telescopes.

Floating platen. A plate located between the main head and the press table in a multi-opening press, and capable of being moved independently.

Force. The male half of a mold, which enters the cavity, exerting pressure on the plastic and causing it to flow; either part of a compression mold.

Gate. In injection molding and in transfer molding, the narrow orifice through which material is injected into a mold cavity from the feed; the molded material removed from the orifice in the process of extracting the mold.

Guide pin. A pin which guides mold halves into alignment on closing.

Head. The end section of a blow-molding machine in which the melt is transformed into a hollow parison.

High-pressure molding. A molding process in which the pressure used is greater than 1000 psi.

Hob. A master model used to sink the shape of a mold into a soft steel block.

Hob punch. The hardened-steel master tool used in hobbing; the male part of a mold.

Injection blow molding. A blow molding process in which the parison to be blown is formed by injection molding.

Injection molding. The method of forming a plastic to the desired shape by forcing the heat-softened plastic into a relatively cool cavity under pressure.

Injection ram. The ram which applies pressure to the plunger in the process of injection molding or transfer molding.

Insert. An integral part of a plastic molding consisting of metal or other material which may be molded into position or pressed into the molding after the molding is completed.

Jet molding. An injection-molding process in which most of the heat is applied to the material as it passes through the nozzle or jet, rather than in a heating cylinder as in conventional processes.

Knockout. Any part or mechanism of a mold used to eject the molded article.

Laminated molding. A molded plastic article produced by bonding together, under heat and pressure in a mold, layers of resin-impregnated laminating reinforcement. See LAMINATES; REINFORCED PLASTICS.

Land. The surface of an extrusion die parallel to the direction of melt flow; the portion of a mold which separates the flash from the molded article; in a semipositive or flash mold, the horizontal bearing surface.

Land area. The whole of the area of contact, perpendicular to the direction of application of pressure of the seating faces of a mold; those faces which come into contact when the mold is closed.

Landed force. A force with a shoulder which seats on land in a landed positive mold.

Locating ring. A ring that serves to align the nozzle of an injection cylinder with the entrance of the sprue bushing and the mold to the machine platen.

Locking pressure. In injection and transfer molding, the pressure which is applied to the mold to keep it closed (in opposition to the fluid pressure of the compressed molding material).

Loose punch. The male portion of a mold, so constructed that it remains attached to the molding when the press is opened and is removed from the mold with the molding (for the purpose of extraction).

Low-pressure molding. Molding at a relatively uniform pressure, ie, of about 200 psi or less, with or without application of heat from external source.

Mandrel. An insert in the flow channel of a die which converts the flow from a solid cross section to some type of hollow or annular cross section. The outer surface of the mandrel guides the flow of the inner surface of the plastic melt before leaving the discharge end of the die.

Manifold. The distribution or piping system which takes the single-channel flow output of the extruder or injection cylinder and divides it to feed several blow-molding heads or injection nozzles. (The term is used mainly with reference to blow-molding and sometimes to injection-molding equipment.)

Matched metal molding. A process for manufacturing reinforced plastics in which matching male and female metal molds are used (similar to compression molding) to form the part, as opposed to low-pressure laminating or spray-up. See REINFORCED PLASTICS.

Mold. The cavity or matrix into or on which the plastic composition is placed and from which it takes form; the assembly of all the parts that function collectively in the molding process to shape parts or articles by heat and pressure.

Mold insert. A removable part of a mold cavity or force which forms undercut or raised portions of a molded article.

Mold-release agent. A material used to prevent sticking of molded articles in the cavity. See RELEASE AGENTS.

Mold seam. The line on a molded or laminated piece, differing in color or appearance from the general surface, caused by the parting line of the mold.

Mold shrinkage. The immediate shrinkage which a molded part undergoes when it is removed from a mold and cooled to room temperature.

Molding. The shaping of a plastic composition within or on a mold, normally accomplished under heat and pressure; sometimes used to denote the finished part.

Molding compounds. A compounded polymer in a form suitable for molding. See COMPOUNDING.

Molding cycle. The period of time necessary for the complete sequence of operations on a molding press to produce one set of moldings; the sequence of operations necessary to produce one set of moldings.

Molding pressure, compression. The unit pressure applied to the molding material in the mold. The area is calculated from the projected area taken at right angles to the direction of applied force and includes all areas under pressure during complete closing of the mold. The unit pressure is calculated by dividing the total force applied by this projected area.

Molding pressure, injection. The pressure applied to the cross-sectional area of the material cylinder.

Molding pressure, transfer. The pressure applied to the cross-sectional area of the material pot or cylinder.

Movable platen. The large back platen of an injection-molding machine to which the back half of the mold is secured during operation.

Multiple-cavity mold. A mold with two or more mold impressions, that is, a mold that produces more than one molding per molding cycle.

Needle blow. A specific blow-molding technique in which the blowing air is injected into the hollow article through a sharpened hollow needle which pierces the parison.

Nozzle. The hollow, cored, metal nose screwed into the extrusion end of the heating cylinder of an injection machine or a transfer chamber (where this is a separate structure). A nozzle is designed to form, under pressure, a seal between the heating cylinder or the transfer chamber and the mold.

Parison. In blow molding, the hollow plastic tube from which an article is blow molded.

Parting line. The mark on a molding or casting where halves of mold meet in closing.

Pinch-off. In blow molding, a raised edge around the cavity in the mold which seals off the part and separates the excess material as the mold closes around the parison in the blow-molding operation.

Platens. The mounting plates of a press, to which the entire mold assembly is bolted.

Plunger. A ram or piston used for the displacement of fluid or semifluid materials in transfer-, injection-, or extrusion-molding methods; also called piston, pommel, force, force plug, and pot plunger.

Positive mold. A mold designed to apply pressure to a piece being molded with no escape of material. See MOLDS.

Pot. A chamber to hold and heat molding material for a transfer mold.

Pot plunger. A plunger used to force softened molding material into the closed cavity of a transfer mold.

Powder molding. Techniques for producing objects by melting polyethylene powder usually against the inside of a mold, either stationary (slush molding) or rotating (rotational molding). See Polyethylene under ETHYLENE POLYMERS; Rotational Molding, p. 118.

Preform. A preshaped fibrous reinforcement formed by distribution of chopped fibers by air, water flotation, or vacuum over the surface of a perforated screen to the approximate contour and thickness desired in the finished part; also, a preshaped fibrous reinforcement of mat or cloth formed to a desired shape on a mandrel or mock-up prior to being placed in a mold press; also, a compact "pill" formed by compressing premixed material to facilitate handling and control of uniformity of charges for mold loading.

Pressure bag molding. A process for molding reinforced plastics, in which a flexible bag is placed over a layup on the mold, sealed, and clamped in place. Fluid pressure, usually compressed air, is placed against the bag and the part is cured. See BAG MOLDING.

Pressure pads. Reinforcements of hardened steel distributed around the dead areas in the faces of a mold, to help the land absorb the final pressure of closing without collapsing.

Profile. The cross-sectional shape of the extrudate; usually reserved for complex shapes. Also, the extrudate with a complex cross section.

Profile extrusion. An extrusion process that produces an essentially finished extrudate, as against one that produces a coating on wire or an intermediary material or parison.

Ram. A piston or plunger which forces molten polymer through the head and die of an accumulator-type injection- or blow-molding machine.

Ram force. The total load (normally expressed in tons) applied by a ram, and numerically equal to the product of the line pressure and the cross-sectional area of the ram.

Retainer plate. The plate on which demountable pieces, such as mold cavities, ejector pins, guide pins, and bushings, are mounted during molding; usually drilled for steam or water.

Rotational molding (or casting). A method used to make hollow articles from plastisols or powders. The material is charged into a hollow mold capable of being rotated in one or two planes. The hot mold fuses the polymer after the rotation has caused it to cover all surfaces. The mold is then chilled and the product stripped out.

Runner. The secondary feed channel that runs from the inner end of the sprue in an injection or transfer mold to the cavity gate; the piece formed in a secondary feed channel or runner.

Scrap. Any product of a molding operation that is not part of the primary product, including flash, culls, runners, sprues and rejected parts.

Screw plasticating injection molding. A molding technique in which the plastic is converted from pellets to a viscous melt by means of an extruder screw that is an integral part of the molding machine. Machines are either single-stage, in which plastication and injection are done by the same cylinder, or double-stage, in which the material is plasticated in one cylinder and fed to a second for injection into a mold.

Semipositive mold. A combination of the positive and flash-type molds; a mold that allows an amount of excess material to escape when it is closed. It is used where close tolerances are required. See MOLDS.

Separate-pot mold. A type of mold used in that method of transfer molding in which the transfer pot or chamber is separate from the mold.

Shot. The yield from one complete molding cycle, including scrap.

Shot capacity. The maximum weight of material which an injection machine can inject with one forward motion of the ram.

Shrink or sink mark. A dimple-like depression with well-rounded edges, in the surface of a part where it has retracted from the mold.

Slush casting (or molding). A method for casting thermoplastics, in which the resin in liquid form is poured into a hot mold where a viscous skin forms. The excess slush is drained off, the mold is cooled, and the molding stripped out. See CASTING.

Solvent molding. A process for forming thermoplastic articles by dipping a male mold in a solution or dispersion of the resin and drawing off the solvent to leave a layer of plastic film adhering to the mold.

Spider. The membranes supporting a mandrel within the head/die assembly.

Split-cavity blocks. Blocks which, when assembled, contain a cavity for molding articles having undercuts.

Split mold. A mold in which the cavity is formed of two or more components held together by an outer chase. The components are known as splits. See MOLDS.

Sprue. The main feed channel that runs from the outer face of an injection or transfer mold to the mold gate in a single-cavity mold or to the runners in a multiple-cavity mold; the piece formed in the feed channel.

Sprue bushing. In an injection mold, a hardened-steel insert which contains the tapered sprue hole and has a suitable seat for making close contact with the nozzle of the injection cylinder.

Sprue lock. In injection molding, a portion of the plastic composition which is held in the cold slug well by an undercut; used to pull the sprue out of the bushing as the mold opens.

Stripper plate. A plate which strips the molded article from mold pin, force, or cores.

Stroke. The length of ram travel on a press.

Torpedo. A streamlined metal block placed in the path of flow of the plastic in the heating cylinder of extruders and injection-molding machines, and used to spread the plastic into thin layers and force it into contact with the heating areas.

Transfer molding. A method of molding in which the plastic is first softened by heat and pressure in a transfer chamber, and then forced by high pressure through suitable sprues, runners, and gates into the closed mold for final curing.

Undercut. A protuberance or indentation that impedes withdrawal from a two-piece, rigid mold; any such protuberance or indentation, depending on the design of the mold. (Tilting a model in designing a mold for that model may eliminate an apparent undercut.)

Upstroke press. A hydraulic press in which the main ram is situated below the moving table, pressure being applied by an upward movement of the ram.

Vacuum bag molding. A process for molding reinforced plastics in which a sheet of flexible transparent material is placed over the lay-up on the mold and sealed. A vacuum is applied between the sheet and the lay-up. The entrapped air is mechanically worked out of the lay-up and removed by the vacuum, and the part is cured. See BAG MOLDING.

Vent. A shallow channel or minute hole cut in the cavity to allow air to escape as the material enters.

COMPRESSION AND TRANSFER MOLDING

Compression and transfer molding are the two main methods used to produce molded parts from thermosetting plastics. Full understanding of the compression- and transfer-molding processes requires an appreciation of a number of factors, in this case of molding compounds, machines, molds, and the parameters affecting the polymerization process of the resin system itself.

History

If the thermosetting plastics industry has a specific founding date it is probably 1909, when Baekeland first produced phenol–formaldehyde resins. The trade name Bakelite of the Union Carbide Company is derived from Baekeland's name.

The emphasis in the early years of thermosetting plastics was in their chemistry. The early pioneers obviously required some pressing mechanism to generate the pressures of 3000 psi and more required for producing dense thermosetting parts; they also required molds to provide the desired shape of the part, and provisions for heating were necessary. The equipment was doubtless crude and the process cycle times were very long (often as much as one hour or greater), but the possibilities for ingenious uses of molded plastics were obvious.

In 1916, another pioneer, Novotny, developed a method for molding cylindrical printing plates, and secured a patent thereon. Novotny, in effect, developed the technique which later became known as transfer molding. He constructed a form consisting of two half-cylinder outer plates, which went around an inner cylindrical section, with approximately ¼-in. space between the inner and outer cylinders. The inside surface of the half-cylinder outside plates was engraved with the appropriate printing characters prior to molding. Then the two cylinders were attached together, and phenol–formaldehyde resin was introduced under pressure through a small hole in this assembly until it filled the entire open space between the two cylinders. Following heat and a suitable cure time, the two half-cylinder plates were removed and a printing roll emerged. This process, while technologically effective and interesting, did not succeed commercially.

In 1926, Shaw obtained patents on the process called transfer molding, and one of the first major commercial applications was the long body of an automatic pencil with the molded spiral grooves on the inside wall.

One of the earliest patents on fully automatic molding presses, issued in the mid-1930s to Zelov, covered an automatic compression machine. Hundreds of these

Fig. 1. Typical compression- and transfer-molded parts.

machines, with subsequent modifications, were introduced to the plastics industry, and some are still in use today.

Some typical compression- and transfer-molded parts are shown in Figure 1.

The Molding Process

The process of compression or transfer molding is carried out in the following manner: A thermosetting molding compound is exposed to sufficient heat, generally approximately 300°F, to soften or plasticize it sufficiently for it to flow into the mold cavity. The fluid plastic is held under pressure, often ranging upwards from 2000 psi, for a sufficient length of time for the material to undergo polymerization or crosslinking, which renders it hard and rigid.

Compression Molding. Historically the most popular molding technique has been compression molding. In its simplest form, compression molding consists of placing a quantity of thermosetting molding compound in the bottom half of an open heated mold which is mounted in a press. The press is then closed, bringing the bottom half of the mold against the top half under pressure. As the mold halves come together, the plastic material begins to soften under the heat, and is formed by the pressure of the two mold halves. After an interval of time which may be a minute or more, depending on the material and thickness of part, the press may be opened, and the plastic will be set in the shape of the mold cavity. The operation of a simple flash mold is shown in Figure 2. More complex types of compression molds are discussed in the article MOLDING.

Transfer (Plunger) Molding. In some instances, it is desirable first to close the mold and then to introduce the molding compound in its fluid state through a small opening or gate leading to the mold cavity. This technique is called transfer

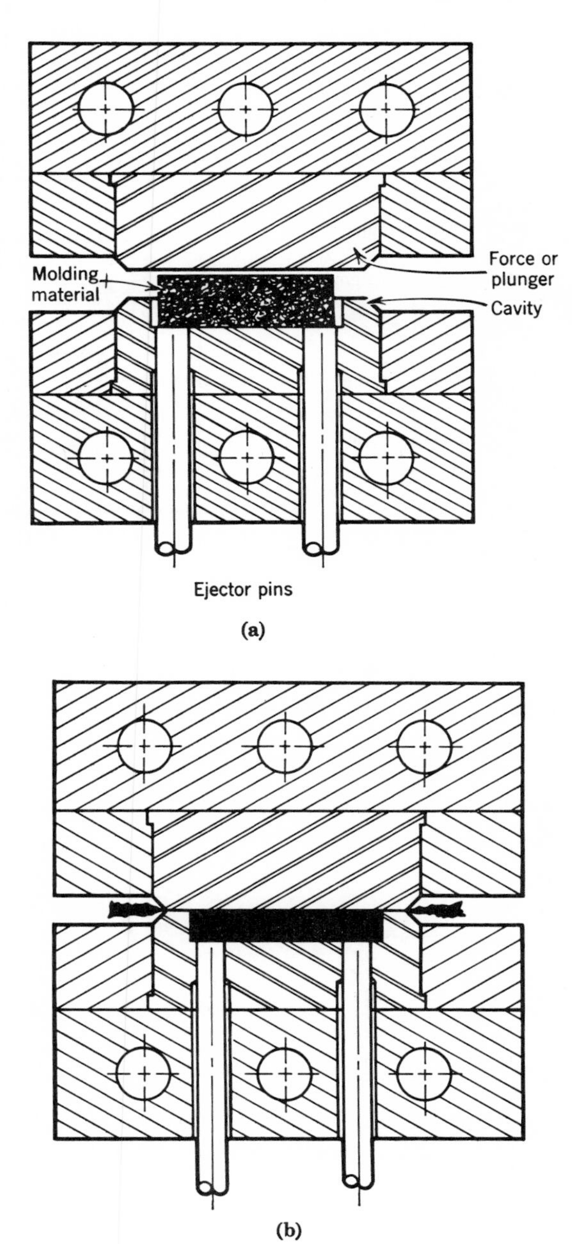

Fig. 2. Operation of a typical flash mold. (**a**) Flash mold open, preform in place; (**b**) flash mold closed.

or plunger molding. It is used frequently when mold sections are very delicate, or when an insert is lightly restrained in the cavity for molding in place. In such applications, closing the mold with the granular compound, as in compression molding, might damage the delicate sections or displace the insert. When the material enters in a softened state, injection speed and pressure can be controlled to minimize the possibility of such damage.

True Transfer and Plunger Molding. In *true* transfer or *pot type* transfer molding (Fig. 3) the mold is closed, then the charge of molding compound is introduced

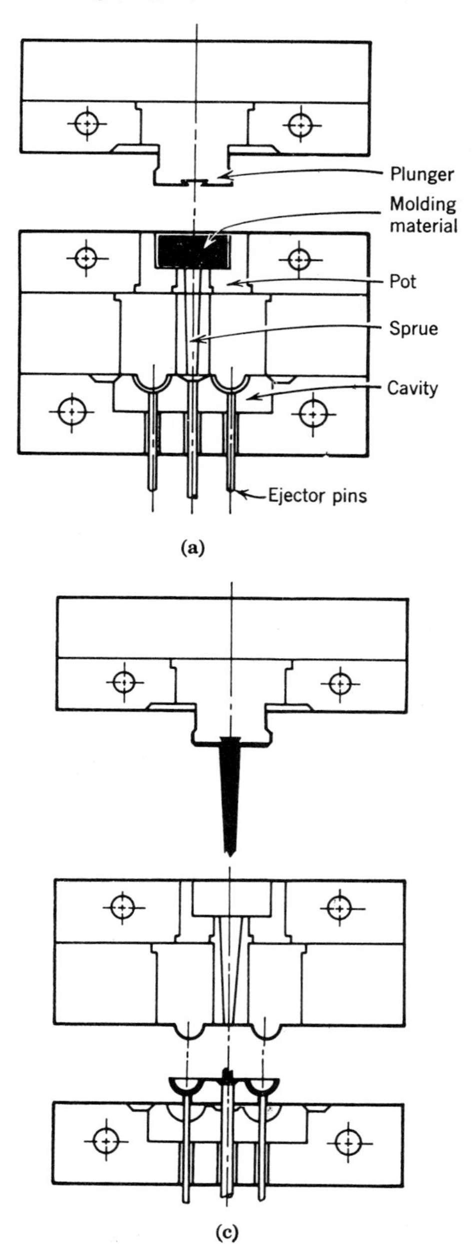

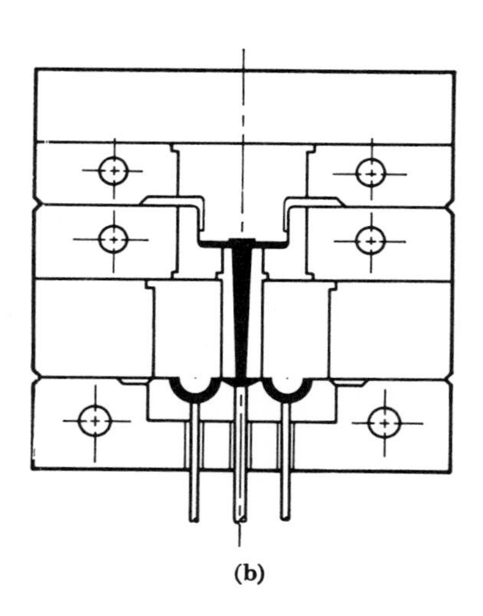

Fig. 3. True, or pot-type, transfer molding. (**a**) Transfer mold (open), pot loaded; (**b**) transfer mold closed; (**c**) transfer mold open, parts ejected, sprue on force plug.

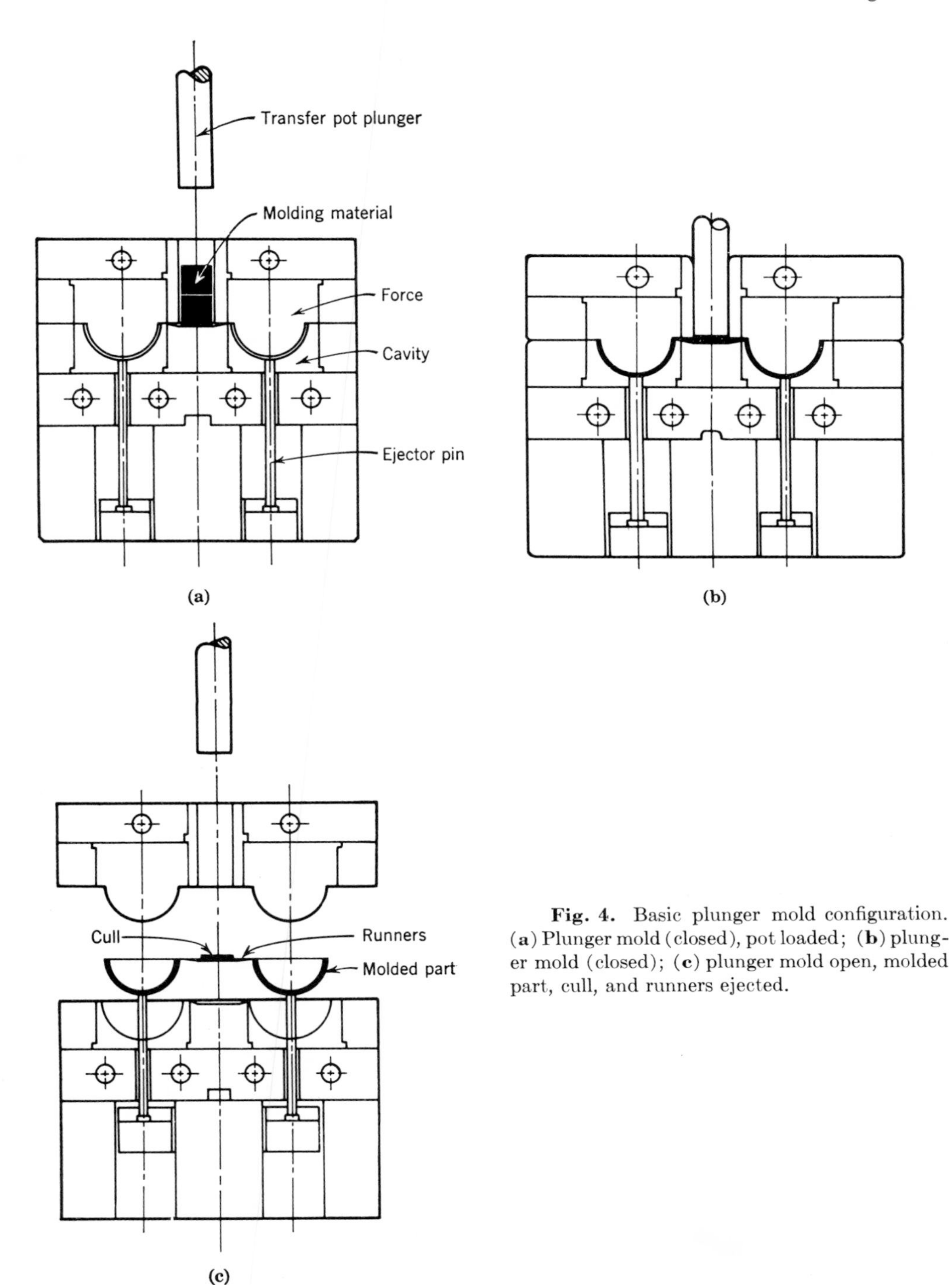

Fig. 4. Basic plunger mold configuration. (**a**) Plunger mold (closed), pot loaded; (**b**) plunger mold (closed); (**c**) plunger mold open, molded part, cull, and runners ejected.

into an open pot at the top of the mold. The plunger is placed into the pot, and the press is closed. As the press closes, it pushes against the plunger, which in turn exerts pressure on the molding compound, forcing it down through a vertical sprue and

Table 1. Comparison of Compression and Transfer (Plunger) Molding

	Compression molding	Transfer molding
press costs	lower	
mold costs		lower
mold wear	higher in cavity section	high in gate area
use of delicate mold sections		more advantageous
cycle times		
semiautomatic, no preheat	faster	
semiautomatic, preheat	comparable	comparable
automatic with preheat		faster
molded-part configuration and properties		
molded-in inserts		better
delicate part section		better
through-holes		better
warpage	less	
strength with medium- or high-impact material	better	
surface finish	comparable	comparable
electrical properties	comparable	comparable
flash	comparable	comparable
wastage (flash, runners, cull)	less	
number of cavities in given die space	greater	
loading time, semiautomatic		less for multi-cavity molds

through runners and gates into the cavities. The area of the pot is about 15% greater than the total projected area of the molded parts and runners, so that the mold will remain closed during this pressing and curing stroke. Following the cure, the plunger is withdrawn, the mold is opened and the parts are ejected.

In the process known as *plunger* molding, the plunger is essentially a part of the press rather than part of the mold. It is usually driven by a hydraulic circuit and a cylinder attached to the head of the press (Fig. 4). The behavior of the molding compound is identical, however, and the decision whether to use true transfer or plunger molding depends on the equipment available and the type of die desired.

In this article the term transfer molding will refer essentially to plunger molding, the most common type of molding in use today. The advantages of compression and of transfer molding are compared in Table 1.

Cold Molding. Cold molding, a variation of compression molding, is a technique of compacting, in a press, usually at room temperature, a compound consisting essentially of a high percentage of filler with a thermosetting binder, then removing the compacted part from the cavities of a mold and placing it in an oven. Under the oven temperature, the binder is crosslinked and effectively holds together the filler material to make a rigid part. Such a process is still used today for making certain high-temperature "porcelain" or "ceramic" parts used in the electrical industries.

Molding of Thermoplastics and Elastomers. Although the vast majority of applications of compression and transfer molding involve true thermosetting plastics such as the phenol–, urea–, and melamine–formaldehyde resins, the alkyd resins, diallyl phthalate polymers, epoxy resins, and silicone compounds, the process is also used with some poly(vinyl chlorides), with rubber, both natural and synthetic, and with a few synthetic elastomers. Such materials can often be processed with either compression-molding or injection-molding techniques, and the selection of technique depends on the part size and configuration.

Elements of the Process. In discussing various parts of any process, it must not be forgotten that the process is, in effect, the combination of several steps or details. In compression and transfer molding, the following five major elements make up the process.

Part Design. The design of the part itself must take into consideration the behavior of thermosetting plastic while in the fluid or plasticized stage, the strength of the material following cure, and aging characteristics. Part sections cannot be too thin, sharp corners should be avoided, flat surfaces should be appropriately ribbed on the backside to minimize warpage, etc. The part structure should be strong enough to withstand the ejection forces at the completion of each cycle. Failure to consider these and other aspects of part design can render a process difficult if not impossible, and certainly can prevent satisfactory economics in molding.

Material Selection. Great care should be taken in choosing the specific thermosetting compound to be used. Various compounds have a wide range of flow characteristics, cure times, and ultimate mechanical, chemical, and electrical properties. The proper combination of end properties and molding properties must be considered to ensure a smoothly operating process.

Mold Design and Construction. Although the basic shape of the mold cavity is determined by the part design, a wide number of other factors affect the design and construction of the mold: (a) The mold must be strong enough to withstand the pressures at the molding temperatures. (b) Provisions must be made for ejecting the molded part. (c) Mold surfaces should be smooth enough to ensure ease of part removal following the molding cycle, and should resist erosion following many cycles. (d) The mold must be designed to facilitate uniform heating to the appropriate temperature level for the material selected. (e) Construction tolerances must be held sufficiently close to prevent flashing or unwanted leakage of plastic material See also Molds.

Machine. Molding machines may be fast or slow, large or small, automatic or manual, hydraulically or mechanically actuated, etc, but for any given production requirement, and for a particular part and mold, usually one type of machine proves ideal. It may be a fully automatic compression machine of 75-ton capacity, or it may be a simple manually operated hydraulic bench press. Selection of the optimum machine is necessary for optimum economics in production.

Operator. Because compression and transfer molding deal with an irreversible chemical reaction, it is vital that the mechanical elements (mold, machine, compound) be properly utilized, and that the many cycle variables (time, temperature, pressure, etc) be regulated suitably. The operator must be thoroughly trained and experienced to contribute this fifth vital element to the process.

Processing Variables

A number of variables, many of them interdependent, play significant parts in the compression- and transfer-molding process.

Molding Compounds. Molding compounds by themselves account for the greatest number of variables. There are thousands of formulations used in compression and transfer molding. The various formulations are often categorized either by their end-use characteristics or by their processing characters. The former include, for example, electrical-grade materials with superior electrical characteristics, medium- or high-impact materials with great tensile and flexural strength, and heat-resistant materials capable of resisting deflection under mechanical load at elevated temperatures. Materials with special processing characteristics include those curing at low temperatures, soft-flowing or low-pressure compounds, fast-curing compounds, etc.

From a process point of view, the viscosity–time curve of any thermosetting compound is often the most critical characteristic. Most molding compounds are granular at room temperature. When exposed to heat, the granules melt and the compound becomes fluid. Under continued heat crosslinking occurs and the material solidifies. As the material goes from solid to fluid to solid, its viscosity changes in a manner indicated by the typical curves in Figure 5. Generally, at some optimum temperature for a given part in a given mold, the viscosity of the fluid plastic will be at a minimum for the necessary period of time to ensure filling of cavity or cavities at an acceptable pressure. If the temperature is too low the material will not achieve this

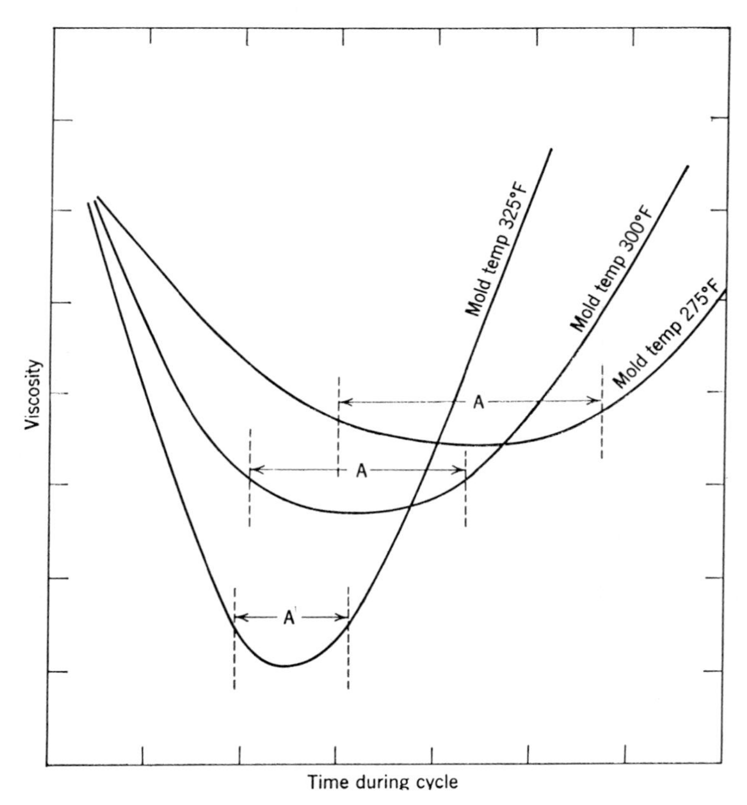

Fig. 5. Types of viscosity–time curves for a typical thermosetting molding compound during flow and crosslinking at different mold temperatures. Region A represents optimum time and viscosity for filling cavity.

low viscosity; if the temperature is too high all the material may not reach this low viscosity at the same time, and may therefore have an average viscosity throughout the particular charge of material which is actually greater than the minimum viscosity obtainable at optimum temperature.

Ideally, in any particular molding operation, the cavities should be filled during this state of lowest viscosity to ensure thorough filling of even the smallest sections of a cavity, a good finish on the molded part, and relatively stressfree moldings. Inasmuch as the viscosity–time curves of molding compounds vary widely, that of the individual molding compound must be known for the particular conditions being used.

Temperature. Most thermosetting molding compounds will crosslink over a wide range of temperatures; however at room temperature, several years might theoretically be required. Some compounds, such as those of B-stage epoxy resins, may have to be stored under refrigeration to retard the tendency to polymerize before molding. The majority of thermosetting compounds must be heated to approximately 300°F for optimum cure. Higher temperatures may degrade some of the physical or electrical characteristics of the material, and, particularly in transfer molding, may cause the material to set up before the cavity is completely filled. Lower temperatures require a longer cure time, thus reducing the productivity of the cycle. There is generally an optimum temperature which produces the best flow characteristics for the particular material and cavity. See also articles on particular polymers, for example, AMINO RESINS; EPOXY RESINS; PHENOLIC RESINS.

The mold is generally heated by electricity, steam, or a hot circulating heat-transfer fluid, to approximately 300°F, the exact temperature depending on the compound and the mold. When the material is in the cavity, heat is transferred from the cavity surface to the molding compound to effect curing.

Because plastics are generally good heat-insulating materials, preheating of the charge is often used to shorten the time in the mold. A compacted charge of the molding compound is heated, either in granular form or in a preform, to a temperature which permits relatively easy handling of the material; the temperature rise is more rapid than can be produced by the mold cavity itself. Preheating may be accomplished by a warm surface plate, by infrared lamps, by hot-air circulation through the granules, by a small oven, or by high-frequency electronic preheaters (see DIELECTRIC HEATING). Preheat temperatures are commonly between 150 and 300°F; higher temperatures are possible when heat is more rapidly transferred to the material.

If the molding compound is held at an elevated temperature for too long during preheating it may cure before it reaches the mold cavity. High-frequency preheating generally proves to be the most rapid method. Preheated material generally flows more readily during the actual molding process, and, because the material starts at an elevated temperature, the time to complete the cure in the mold cavity is shortened, generally yielding a more economical overall cycle.

A third source of heat input to the plastic during the molding process is from frictional heat during the closing of the mold. In the case of compression molding, the material is forced to flow by the closing action of the mold. This flow may be at fairly high localized velocities, which impart a certain amount of frictional heat energy to the plastic. In transfer molding, the frictional heat is even more pronounced as the material is forced along runners and through relatively small gates leading to the cavities. The amount of frictional heat added to the material in transfer molding is dependent on the speed of plunger advance, the size and configuration of runners and gates, and the surface finish of the mold. It must be taken into consideration in setting up an overall process, particularly when using relatively heat-sensitive materials.

The fourth and final heat input is during postcure. With many thermosetting compounds, an exposure to elevated temperatures following removal of the molded part from the mold increases the electrical and mechanical properties of the plastic. A typical postcure, for example, might involve 2 hr at 275°F, followed by 2 hr at 150°F. Generally, such postcuring produces, under controlled conditions, the ultimate crosslinking of the compound. Without postcure, the molded part may undergo the final 5 or 10% of crosslinking over a period of months or even years, particularly in the common phenol–formaldehyde resins. In general, if no postcure is contemplated, the part is left in the cavity slightly longer to ensure an optimum degree of cure. However, when postcuring is to be used, the cure time in the mold may be slightly shortened, to yield optimum efficiency in the use of machine and mold.

Time. Once the temperature has been determined at which the molding compound will properly fill the cavities without precure and without burning of the part, the correct cure time is established by reducing the time to the point where the molded part shows blisters, indicating undercure, then increasing the time slightly to allow a safe margin. In high-production operations the difference of a few seconds in cure time for any single cycle may affect the total production by 5 or 10%. During the preheating cycle, time control is critical to ensure that each charge of molding compound will be preheated to essentially the same temperature so that consistent results are obtained during the actual molding and in the end properties of the part.

Breathe and Dwell. Some thermosetting molding compounds, such as the phenolic resins, give off gaseous products as part of the polymerization reaction. Under high pressure these gases can be contained in the molded part until curing is fairly complete; the part then may be removed from the cavity with no visible change. If the part is held in the cavity for too short a time, however, after it is removed it will probably display blisters or perhaps large bulges or even ruptures caused by these contained gases, since the resin has not crosslinked sufficiently to retain its shape against the force of their pressure. With such resin systems, particularly in compression molding, it has been found possible to release a large quantity of these gases before the polymerization is completed. To release the gases, the mold is allowed to open at some set time following close to allow the cavity to "breathe" or "degas." The length of time that the mold remains partially open for the breathe cycle is called the "dwell" and depends on the size and configuration of the part. A breathe-and-dwell step during the actual molding cycle will eliminate most of these gases. It will also permit a shorter cure time because the resin will crosslink sufficiently in a shorter length of time to withstand the lesser pressure of less contained gases. A typical compression-molding cycle might proceed as shown in Figure 6. Obviously, timing is important in starting the breathe cycle when the material has reached an optimum point of gas generation, and in holding the mold open long enough for much of the gas to escape; however, the mold must not be held open so long that the material will cure while the mold is partially open. With some parts and some compounds a double breathe operation is used.

It is well to note that with materials giving off such a gaseous reaction product, even with a breathe cycle the final molded part will probably exhibit outgassing tendencies for variable lengths of time after molding, even for months and years. In selecting

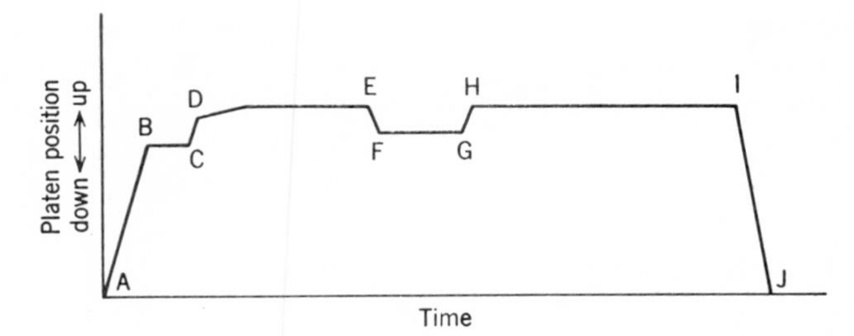

Fig. 6. Schematic diagram of positions of platen during a typical compression-molding cycle with breathe step. Key: A, start of cycle; A–B, platen ram raised at high speed; B–C, slow closing until contact with molding powder; C–D, rapid closing with compacting of molding powder; D–E, pressing; E–F, press opening for breathing; F–G, dwell for breathing; G–H, mold closing; H–I, curing; I–J, press opening; J, end of cycle.

molding compounds, outgassing in applications under vacuum, or in space, or in the vicinity of critical electrical contacts must be taken into consideration. In addition, if molded parts are to be metallized or plated, outgassing tendencies can actually rupture the plated film.

It should be pointed out, furthermore, that even when there is no gas of reaction, a breathe cycle may be indicated for materials which might have absorbed water vapor from the air prior to molding. This moisture turns to steam at the elevated temperatures of the molding cycle, and it may be desirable to release this steam by a breathe step in the molding cycle.

In connection with the presence of moisture in molding compound, it should also be pointed out that on occasion, particularly in years past, molders would actually add a small quantity of moisture to molding compound in an effort to speed up the cure cycle. The quantity of moisture might be as high as 2–3%. The theory was that the moisture would turn to steam, which would permeate the molding compound in the cavity causing a better heat transfer to all of the compound, and thus a faster cure. This technique of adding moisture is rarely used in modern plants with up-to-date preheating equipment.

Breathe-and-dwell cycles are rarely used in transfer molding. Obviously, before opening the mold for the degassing or breathing step, it is important that the pressure on the transfer plunger be removed. Following the breathe-and-dwell steps, the mold is again closed and pressure is restored to the transfer plunger.

Pressure. Although curing reactions will take place at atmospheric pressure, molding with thermosetting plastics (as opposed to casting with thermosetting plastics) requires greater pressure for two basic reasons: (a) to ensure that the plastic fills all of the cavity and has relatively uniform density throughout (pressure causes the cavity to fill, and resists any tendency of internal gases to form voids or gas pockets; pressure must of course be sufficient to overcome resistance of the plastic to flow); (b) to ensure better heat transfer to the material (higher pressure produces a higher density, which generally means faster thermoconductivity).

Pressures needed for molding phenol–, urea–, and melamine–formaldehyde resins as well as many other conventional thermosetting resins generally range from 3,000 to 10,000 psi. A number of low-pressure molding and encapsulating compounds may mold well at pressures in the range of 100 to 1,000 psi. In transfer molding, although pressures in the cavity may be in the above ranges, it may be necessary to exert considerably higher pressures in the transfer pot to ensure that the material will flow rapidly to the cavities and that any reduction in pressure between the plunger head and the cavities themselves will still leave the necessary pressure in the cavities. Under pressure, molding compounds behave as non-Newtonian fluids, and therefore do not have the same "hydrostatic" pressures at the plunger head as in the cavities. The degree of pressure drop during actual transfer molding depends on the molding-compound characteristics as well as on the layout of mold cavities, gates, and runners.

In compression molding, adequate pressure is needed to ensure a good "pinch-off" and a thin parting line. In transfer molding, excess pressure may result in mold flashing, or in damage to delicate inserts in the cavity. Because of the effect of pressure in these areas of the molding cycle, pressure control is important. In modern equipment pressure can be regulated to within 5% of the desired value.

Part Size and Design. The design of the molded part will affect the molding cycle. For example, since heat transfer to the interior of a thick part takes time, a part with thick cross sections will generally have a longer cure time than a part with thin sections, particularly in compression molding. If an entire part is made of thin sections, and there is one thick section, the cure time will be dependent on that one thick section. If possible, therefore, thick sections should be replaced by suitable ribs or bosses or be cored out on the backside. In transfer molding, part thickness is not quite as important because the material absorbs heat as it is pushed down runners and through gates, where every particle is close to, or touching, the hot surface of the mold.

Tolerances in part design can also affect the length of the cure. If tolerances are extremely close, cure must often be extra-long to produce a greater degree of cure, so that after the part is removed from the cavity there is less chance for warpage to occur during final curing. Dimensional stability, in other words, is often dependent on the degree of cure. If tolerances can be relaxed, the part may be removed from the cavity after a shorter cure and production costs are thus reduced.

If a part is fairly deep in the cavity, a draft or taper should be allowed so that it can be pushed out of the cavity readily. Without a draft, the ejection force may have to be greater, requiring longer cure before ejection.

Speeds. To achieve a fast molding cycle, high speeds during each step are not necessarily advisable. More important is the control of speed in certain steps of the cycle. The cavity should be filled rapidly, but at a rate commensurate with the viscosity–time curves of the molding compound at the prevailing temperature. In compression molding, for example, if the mold is closed too rapidly it will force the cavity sections to bear on the molding compound, which is not yet fully plasticized. This will cause extreme local pressures which could abrade the mold surface, damage fragile mold cavity sections, and trap air in the molded part. During this closing operation in compression molding it is advisable to have an accurate temperature-compensated speed-control valve on the hydraulic circuit serving the clamp ram.

The speed should be so regulated that the material is kept in relatively continuous flow, and that when the mold is finally reaching the fully closed position, enough heat has been transferred to the molding compound that it flows relatively freely. The actual time to effect the final ¼ in. of closing may be 2–10 sec or more. Under certain conditions it may even be advisable to arrest the closing of the mold for a number of seconds while more heat is transferred into the material, and then to complete the closing.

At the end of a compression-molding cycle it may be advisable to have a controlled speed of "break-away" or first opening, to ensure that the plastic part will not be damaged from too violent a separation of the mold halves. At the end of the opening stroke on either compression or transfer molding the ejection of the part should be at a controlled speed, such as to minimize possible damage to the part from too violent ejection.

In transfer molding, the material should travel down the runner and through the gates into the cavity in the shortest possible time to ensure uniform density and uniform curing. If the material moves too rapidly, however, it may absorb too much frictional heat, causing it to set up prematurely. If there is too much speed through the gate, the material will be highly turbulent as it enters the cavity, and may tend to trap air, leading to voids in the finished part. Also, the force of the plastic entering the cavity rapidly may dislodge or break delicate inserts. A controlled speed of plunger movement ensures that the material will move smoothly through the runners and gates and will fill the cavities gently, with sufficient time allowed for the escape of air or gases through the cavity vent.

The movement of the platen during mold closing, as well as that of the transfer plunger, should generally be rapid until the material begins to flow, then should be slower and accurately regulated during the actual flow (Fig. 7).

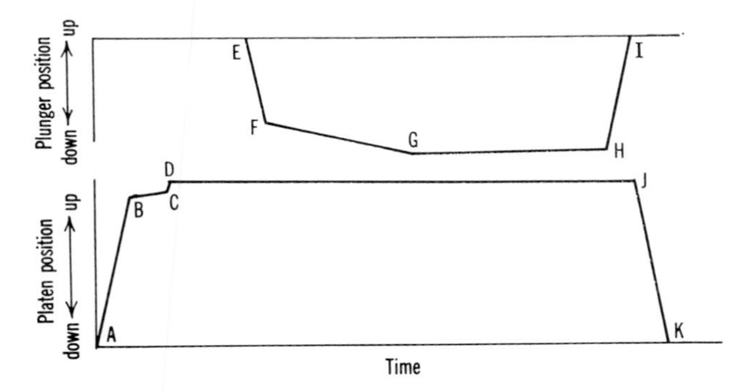

Fig. 7. Schematic diagram of positions of mold platen and of transfer plunger during typical transfer-molding cycle without breathe step. Key: A, start of cycle; A–B, platen ram raised at high speed; B–C, slow closing until contact of force with cavity; C–D, closing of mold; E–F, transfer ram into pot at high speed; F–G, molding powder compacted and transferred to mold; G–H, cure: H–I, transfer ram out of pot; J–K, press opened; K, end of cycle.

Equipment

Equipment necessary for compression and transfer molding falls into three principal categories: the basic machine, auxiliary equipment, and molds (qv). The actual equipment used will depend upon the type of item being produced, its design, the plastic used, and the rate of production. For example, some parts may be produced by the millions each year, whereas only a few thousands or less may be made of others.

Compression-Molding Presses. Compression-molding presses can be discussed in three major categories: the press frame, the driving means, and the controls.

Press Frame. Because compression molding requires bringing two halves of a mold together under pressure, the basic press frame has two or more flat steel or cast-iron slabs, called platens, which move toward one another to effect the pressing stroke. Traditionally, on a vertical press frame, only one of the platens moves up or down. The platens are supported or guided by one or more means: (a) two or more tie rods or side rods; (b) slab sides, with bearing surfaces; or (c) a rigid C-clamp frame, referred to as an arbor press.

When the moving platen moves upward toward a fixed upper platen, the press is referred to as an upward-acting press. When the pressing stroke is effected by the downward movement of one platen toward a lower platen, the press is referred to as a downward-acting press. Multiple-platen presses, with three or more platens, can accommodate several molds at the same time.

The majority of compression and transfer presses are vertical; however, a number of the more modern automatic units are horizontal in configuration, with platens moving left to right. In effect, a horizontal press is merely a vertical press frame tipped over on its side. Rotary presses are also available. These have several pressing stations or presses positioned around a circular table.

In all of these configurations the frame and platens must have enough strength to minimize deflections even after many years of use.

Driving Means. Several possibilities exist for forcibly moving the platens on the press frame toward one another and apart. The most common means is the hydraulic ram. (Figure 8 illustrates a simple vertical hydraulic ram press.) The typical compression press today has a double-acting hydraulic ram (that is, it uses high-pressure oil both to move the platens together and also to move them apart) driven by a hydraulic pump which, in turn, is driven by an electric motor. The pumps generally operate at 2000 or 3000 psi, and are driven by electric motors at either 1200 or 1800 rpm. Lower-pressure pumps may be used, with hydraulic boosters adding the high pressure.

The driving means on such a machine consists of the hydraulic reservoir, pump, motor, necessary valving and piping, and the hydraulic cylinder itself. The cylinder is mounted on the press frame and is, of course, attached in such a way that the ram drives the moving platen up and down (or forward and back). Hydraulic power units can be fairly sophisticated, and generally have a multiplicity of check valves, sequence valves, throttling valves, etc, to produce the desired speeds and pressures required for a modern molding cycle. On larger presses, jack rams are sometimes used to provide high-speed movement "in the clear" without excessive pump capacities. On the presses of below 100-ton capacity, so-called regeneration cycles are frequently used for high speeds in the clear with minimum pump capacity. In many plants where there are many presses a hydraulic accumulator is used instead of individual pumps and reservoirs.

Presses of up to 25-ton clamping capacity may use air cylinders and air pressure instead of hydraulic systems. Combinations of air and oil are occasionally used for economy of operation. In such installations, shop air pressure is used in conjunction with an oil intensifier working on an otherwise static oil system to achieve high pressures in the clamp. Another system utilizes a hydraulic cylinder and ram in conjunction with a mechanical toggle to provide mechanical advantage for the clamp stroke. Hydraulically actuated toggle mechanisms often produce ten or even twenty times the clamping force that would be obtained by the hydraulic cylinder only.

Another driving means is the screw-type drive, in which an electric motor rotates

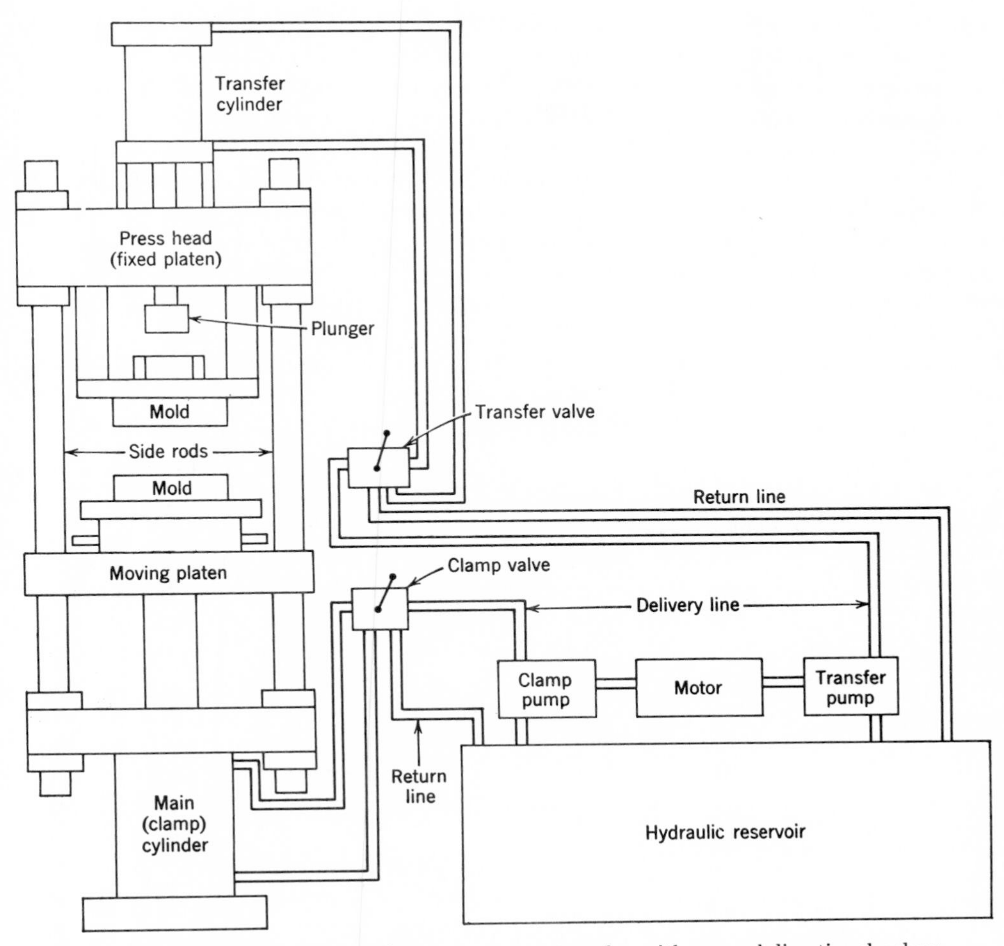

Fig. 8. Simple vertical hydraulic ram press, top transfer, with manual directional valves.

a large threaded shaft or lead screw. A nut advancing on the shaft closes the platen and applies the necessary clamping force. Screw presses are often very slow.

The driving of rotary presses is often effected by the rotation of the large rotary table on which are mounted the several presses. As the table rotates, an extension on the moving platen of each press rides over a fixed spring-loaded cam causing the mold to close. With continued rotation of the table the mold is held closed until the extension finally rides off the cam, permitting the mold to open. Individual press stations in a rotary press may be limited to 3, 5, or possibly 10 tons, whereas large vertical hydraulic ram presses may produce as much as 2000 or more tons of clamping force.

Control System. The control system actuates the driving means to effect the necessary platen movements for molding. The simplest control system is the manual control, in which the operator manually moves a lever that positions a spool in a valve to cause a hydraulic or air press to close or open. The valve generally has three positions, ie, close, open, and neutral, and the operator can generally effect throttling control by easing the valve lever into the "close" or "open" position.

Semiautomatic controls are becoming increasingly popular, as they minimize the human error during a molding cycle. The operator must push electrical contact buttons to initiate a cycle. The electrical control system then causes a four-way valve to shift to close the press; timers and other devices employing limit switches cause the press to effect slow-close, full-close, breathe-open, breathe-dwell, close, cure, and press-open. Such semiautomatic control systems can be quite sophisticated; they may include additional safety interlocks to prevent malfunction and to permit the operator to use manual control in an emergency. They must be reactuated each cycle by the operator.

Fully automatic controls are required for fully automatic presses and are discussed later.

Transfer Presses. In transfer, or plunger, molding, the mold halves are brought together under pressure, as in compression molding. The charge of molding compound is then placed into a pot and driven from the pot through runners and gates into the mold cavities by means of a plunger.

On modern transfer presses the ram is mounted on the press frame. The transfer ram is generally mounted on top of an upward-acting press frame; such a press is referred to as a top-transfer press. Conversely, in a downward-acting press, the transfer ram is generally mounted on the bottom fixed platen, and the press is referred to as a bottom-transfer press.

Transfer presses can be horizontal. Occasionally the transfer ram is so positioned that it moves in the plane of the mold parting surface, moving in a direction perpendicular to the press platen direction. The driving means for the transfer ram is generally hydraulic, but may be air. Air–oil, toggle, or screw drives are rarely utilized for transfer-ram operation.

For transfer presses, the control systems include the same features as for compression presses. In addition, however, they generally require the operator to actuate the transfer cycle after he has placed the charge into the pot. A typical sequence of control operations in a transfer-molding process is as follows: With the press and mold open and the transfer ram retracted the operator places the inserts (if required) into the open cavities; the operator then actuates pushbuttons which cause the press to close and hold the mold closed under pressure; the operator places the molding compound into the transfer pot and actuates pushbuttons which cause the transfer plunger to enter the pot under pressure; semiautomatic controls ensure that the plunger holds pressure for a set cure time, then cause the press to open and the transfer ram to retract; semiautomatic controls automatically reset the timers and place all valves in neutral position; the operator removes the finished parts, cull, and runners, blows off mold surfaces, pot, and plunger, and starts the next cycle.

Accessories. The major accessories used in compression and transfer molding are the heating and the ejector mechanisms. Minor accessories basic to the press movement often include a transfer-speed control, in which the transfer ram, on a semiautomatic cycle, moves rapidly until the plunger actually meets the charge of molding compound. It then slows to a controlled speed and pressure for the actual transfer phase, and holds the set pressure for the duration of the cure. A platen "slow-close" is an accessory on some machines; by this means the press platen can be made to slow to a desired speed at a certain point in the closing stroke.

Heating. Compression and transfer molding with thermosetting plastics requires the application of heat, as well as pressure and time, to bring about curing. Provisions for heating the mold are therefore basic in any compression- or transfer-press installation.

Three common types of mold heating are utilized. High-pressure steam capable of producing mold temperatures up to approximately 340°F is used in many large custom-molding plants. Steam generators are necessary, either of the centralized high-pressure boiler type, or of the smaller self-contained gas or electric types which can serve one or two presses only. Steam provides a good even heat, and temperature control, effected by steam-pressure regulators, is very uniform. Steam lines often prove to be obstacles around a press or mold, and steam leaks are generally routine.

Electric heating, with cylindrical heating cartridges or flat heating strips mounted against or inside the heating platens or molds, is becoming increasingly popular. Molding can be carried out at temperatures approaching, or even exceeding, 400°F (if the mold and molding compound will withstand such high molding temperatures); faster cures result. Electric heating is generally cleaner than steam, and is fairly simple to install. Temperature controllers are necessary, sometimes of the simple "on-off" type, and sometimes with a more elaborate "proportional" set-up. The more costly and more accurate electrical temperature controllers use a thermocouple-type sensing element, mounted on or in each mold half, to send electrical signals to a potentiometer-type instrument which, in turn, controls the heavy-duty contactor admitting current to the heating elements. Less expensive, but generally satisfactory, are the filled-bulb systems, with a sensing bulb located in each half of the mold, and a capillary tube transmitting the pressure from the heated bulb to a pressure-sensitive meter which, in turn, actuates the heavy-duty contactor.

A third common method of mold heating is the circulating hot-oil system. The molds or heating platens are channelled to permit the flow of a heat-exchange medium. The heat-exchange medium (synthetic oil) passes through an auxiliary unit positioned near the press where heat is provided, generally electrically; sensitive temperature controllers maintain the temperature of the circulating oil at the desired level. The oil circulates continuously to maintain uniform temperature. Circulating-oil heating systems are often used where a cooling cycle following the heating cycle is desired.

Ejector Mechanisms. Another basic accessory generally found on modern compression- and transfer-molding presses is the ejector mechanism. Most molds include a knockout plate to which are connected the knockout pins that eject the molded parts from the mold cavities. Often mounted on the press frame is a means for moving the knockout plate with respect to the mold to eject molded parts at the end of each cycle and then to return the knockout plate for the next molding cycle. Ejection mechanisms may be mechanical, in that they are mechanically actuated as the moving platen reaches its full open position. Mechanical ejection systems can be adjusted for length of stroke, and generally ensure completely parallel movement of the knockout plate. After the molded parts are ejected, it is often necessary to jog the moving platen 1 or 2 in. toward the closed position to retract the knockout plate.

Hydraulic ejector systems offer greater flexibility in that the ejection stroke can take place after the press platen has reached its full open position. Speeds and forces of hydraulic ejector systems are also more closely controlled than those of mechanical ejector systems. An added convenience of hydraulic ejector systems is that they usually may be returned to the retracted position merely by actuating a pushbutton or selector switch. See also MOLDS.

Preheaters. A number of accessories are also found external to the press, power unit, and control installation itself, one of the most important of which is the preheater. Preheating of the molding charge generally makes possible shorter cures, and thus a higher production rate. The most popular method is dielectric heating (qv). Such preheaters are located adjacent to a molding press, and are manually operated by the operator for each cycle. Other types of preheaters use hot air, infrared radiation, or steam as the heating medium. Because most thermosetting molding compounds are good heat insulators, dielectric preheating offers the fastest means of raising the temperature of the molding compound.

Preformer. A preformer is recommended auxiliary equipment for high production rates. A preformer compacts the granular molding compound into charges of uniform size and weight. It is easier for an operator to pick up a preform and place it in a mold cavity or transfer pot than to have to weigh a charge of granular material One preformer can generally serve a number of compression or transfer presses. Preforming is therefore often done in an area somewhat remote from the molding area, with the preforms delivered to the molding installation as required. Preformers are basically compacting presses; they may be mechanical, hydraulic, pneumatic, or rotary cam-type machines.

Flash Removal. In any compression- or transfer-molding installation it is necessary to remove flash from the molded parts. This is most often accomplished with tumbling machines, which, as the name implies, tumble molds parts against each other to break off the flash. The simplest tumblers are merely wire baskets driven by a small electric motor and pulley belt. Tumbling machines, however, can become quite elaborate, involving not only the tumbling operation, but often the "blasting" of molded parts during the tumbling operation, usually with granulated peach or

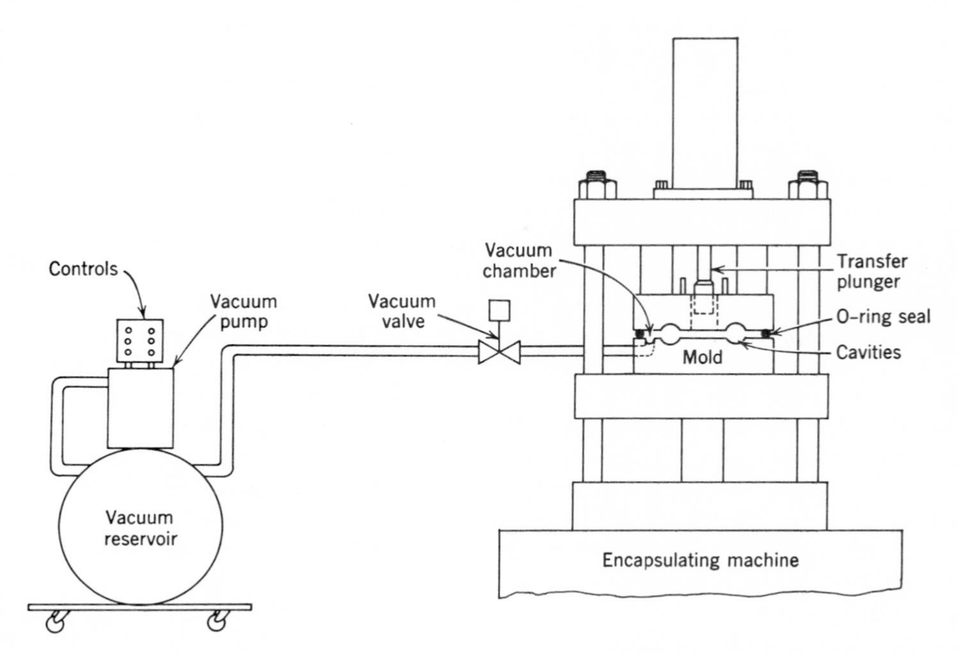

Fig. 9. Vacuum-venting process for encapsulation by transfer molding.

apricot pits. A steam jet is sometimes used during the tumbling operation to minimize accumulation of static charges on the tumbling parts. One large tumbling machine can service many molding presses.

Postcuring Ovens. Some thermosetting molding compounds exhibit improved properties if they are exposed to a postcure of one hour or more at specified temperatures. Occasionally the parts are mounted on shrink fixtures to minimize dimensional change during the postcuring operation.

Vacuum Venting. A vacuum venting system is occasionally required in transfer-molding operations. Its purpose is to remove most of the air from cavities prior to having the plastic enter the cavities; it thereby minimizes the possibility of voids, and often results in shorter cure cycles. Figure 9 shows a schematic diagram of the vacuum system used in a transfer-molding press such as that in Figure 10.

Additional Accessories. Other optional features include a variety of means of affixing molds to press platens: T-slots are most popular on larger presses, but drilled and tapped holes are common on smaller presses. Miscellaneous features include devices to shut the press down if the hydraulic oil becomes too hot; provisions for circulating cooling water through a heat exchanger in the pumping system, to keep hydraulic oils at optimum operating temperatures; oil filters, to prevent damage of hydraulic valves and pumps from any impurities which might develop in the hydraulic system; fans to cool enclosed hydraulic power units; and indicating lights for visual check of the various stages of molding cycle in process.

Fig. 10. A highly instrumented transfer-molding press. At left is a vacuum-venting system, with large vertical vacuum reservoir; between the vacuum-venting system and the press is a hopper-fed preformer.

Types of Tooling. Both hand tools and production tools are used in compression and transfer molding. Hand tools are generally selected for experimental work, and low or moderate production. Figure 11 shows a typical hand tool for experimental work. Such hand tools can often be relatively inexpensive, as befits an experimental mold. On the other hand for most development programs, and certainly for more extended production, hand tools should be made from hardened tool steel, highly polished and chrome plated. Additional costs for such sophisticated tooling generally are offset by the ease of operation during the molding process. Hand tools generally do not have integral heating provisions, but draw heat from heated platens which remain in the molding machine during the process. Hand tools may have knockout pins or other provisions for ejection of molded parts.

Self-contained production tools are generally used for optimum economy at moderate or high production rates. These tools usually have more than one cavity, and generally are bolted into the molding machine and are not removed from the machine during a given production run. The number of cavities in a production

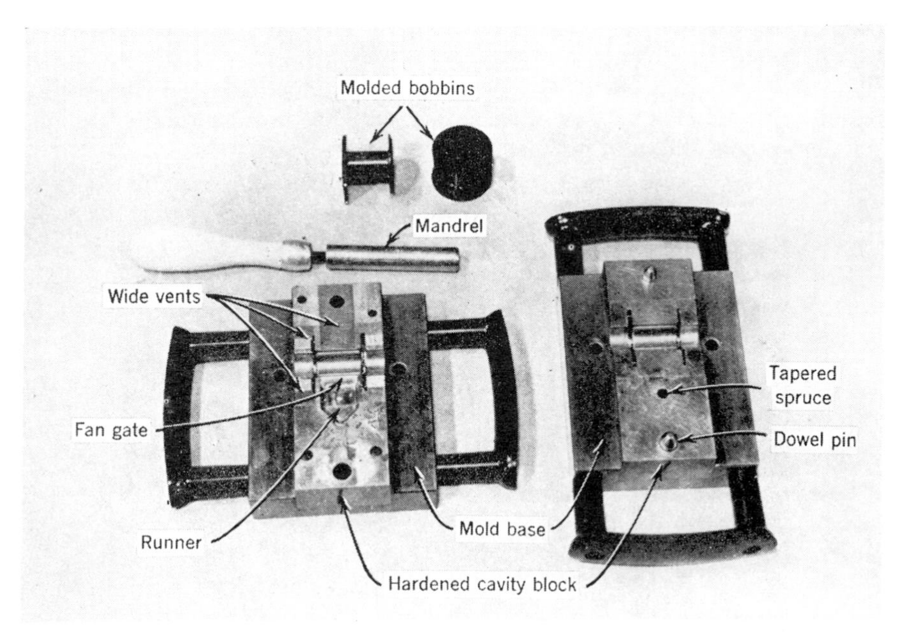

Fig. 11. Hand transfer mold for coil bobbin, to be used with universal heating platens equipped with pot, in a plunger molding press. Courtesy Hull Corp.

mold is dictated by the production rate required, the tonnage capacity of the molding machine, and the efficiency of loading and unloading any inserts which are to be molded in. Frequently, loading fixtures are an accessory of production tools if such inserts are to be used. Figure 12 shows a typical production tool outside the molding press, and Figures 13 and 14 show self-contained production tools mounted in typical molding machines.

A hybrid hand–production tool perhaps with a removable bottom half and a fixed top half, or with other such variations, may prove best for certain parts.

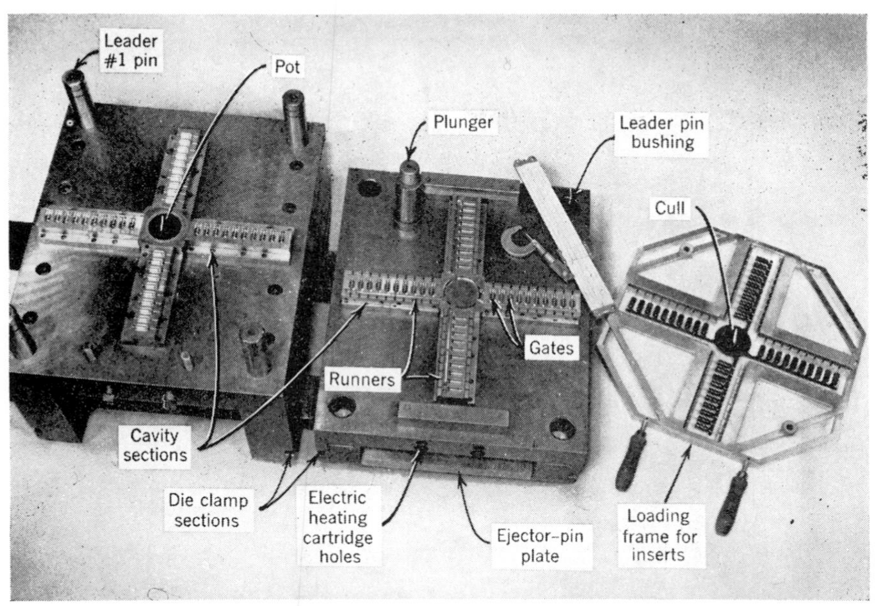

Fig. 12. Transfer mold showing typical elements. Courtesy Hull Corp.

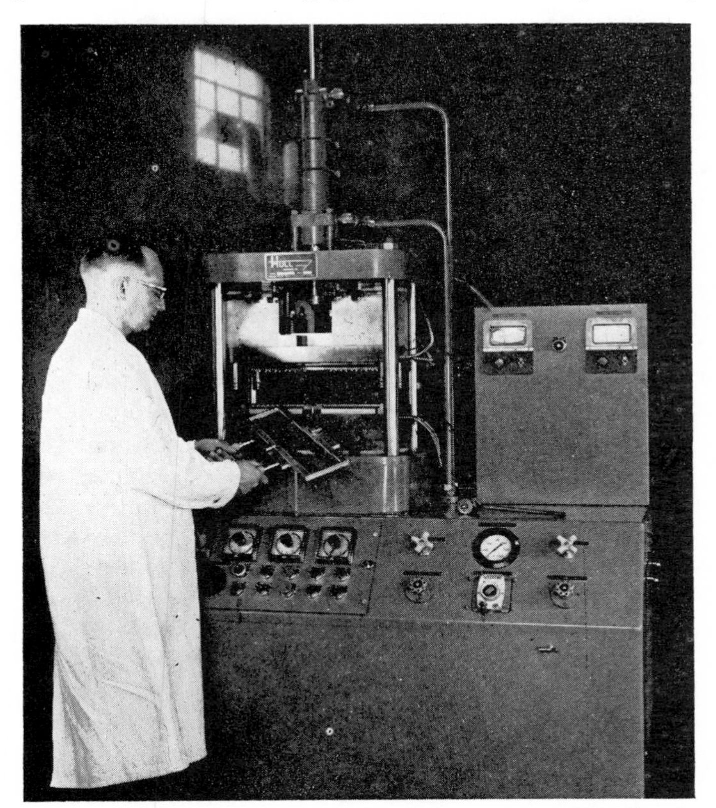

Fig. 13. Production mold mounted in machine, with operator and work-holding fixture to simplify loading of multiple inserts.

Fig. 14. Production mold for molding four thin-walled cases and four covers per cycle. Courtesy Hull Corp.

Cavity Design. The details of the cavity design are, of course, dictated by the part design. The cavity should be so shaped, with curved sections or with some draft, to make possible easy removal of the molded part. Uniform thicknesses of the plastic are desirable to minimize localized stresses from uneven curing of the molding compound. If an insert is to be molded in, the cavity design must provide for supporting such an insert adequately during the mold closing and during plastic flow and curing.

The cavity design should take into consideration the ejection of the part following molding. In hand molds, *ejection pins* are generally not used; often the molded part can be blown from the cavity with an air blast, or can be gently removed by hand. In production molds, ejector pins are almost mandatory at high production rates. Where ejector pins are used, they should be of sufficient diameter, and there should be a sufficient quantity of such pins, that localized stresses at the point where the ejection pin bears on the molded part will not be so severe as to damage the part during ejection. (Inadequate pins may necessitate a longer cure to produce a harder part for ejection.) Additionally, the pins should not be so large or so positioned that they disfigure the product with the inevitable ejector pin mark.

The *gate location* should be such that the material enters the cavity where there are minimum restrictions to further flow. The location may also be selected with a view to minimizing any disadvantage of the gate scar on the finished part. *Gate shape* may be a pin point, a rectangular cross section, a V, a half round, or even a fan shape. Some parts require ring gating or multiple gates for minimum stresses and more uniform cavity filling. Gates may be shaped to produce a clean breakoff right at the surface of the molded part, thus eliminating the need for trimming dies or for further grinding to remove any protruding plastic at the gate. If fiber-reinforced compound is used, and if the runner is broken off right at the body of the molded part, however, some of the fibers actually pull away some of the compound from the body of the part, leaving an indentation. Therefore with such materials the gate is generally designed so that its thinnest part is actually outside the body of the molded part. This arrangement ensures that the runner will break off outside the part; the gate can be finished flush with a grinding operation. Occasionally provisions are made for easy replacement of just the gate section of the cavity. Thus, when gate wear occurs, minimum maintenance is required to restore the mold to new condition.

In both compression and transfer molds it is necessary to provide for escape of air from the cavity as the molding compound fills all of the cavity volume. In compression molding, the breathe cycle may prove the best way to ensure adequate escape of entrapped air. In transfer molding, the incoming compound tends to sweep all air in the cavity toward the side away from the gate. It is thus generally necessary to provide *venting.* Vents are usually machined into the parting line, often at corners or other parts of the cavity shape where air might be entrapped. Vents may be several thousandths of an inch deep, depending on molding compound, and may be as much as 0.5 in. wide, or even wider, depending on the total volume of the part and the amount of air which must escape. Although a toolmaker can often estimate the location and size of vents, only molding in the actual cavity will determine when and where additional vents are needed. Vents should be highly polished and shaped so that molding compound which cures in the vent will readily eject with the molded part, and will not remain behind to clog the vent in successive cycles.

One of the newer techniques for adequate venting of cavities with complex shapes is *vacuum venting,* which is used in transfer molding. It consists in applying vacuum to the cavity after the mold is closed and prior to the introduction of molding compound. Both hand molds and production molds can be designed for use with vacuum venting. Although the actual cavity and its vent are generally quite similar with or without vacuum venting, the rest of the die base is quite different in that it must provide a vacuum-tight seal and have a suitable evacuation chamber around the cavity sections and, of course, it must have a vacuum port which is connected to a vacuum system.

Some molds use hydraulically actuated *positioning pins* to hold an insert in the correct position in a cavity during the time molding compound flows around the insert; these are retracted before the material cures, in order that additional plasticized material may fill the voids where the positioning pins were located. Such positioning pins often serve as ejection pins upon completion of the cycle.

Cavity *dimensions* require consideration of normal mold shrinkage of the particular molding compound being used. Mold shrinkage may run from 0.002 in./in. to as high as 0.020 in./in. Such shrinkage rarely takes place uniformly throughout a molded part. It is generally influenced by local flow conditions in compression and transfer moldings and generally calls for experience in determining its extent.

Some cavities require removable mold inserts for best operation. Removable inserts are advisable, for example, when the finished shape must include undercuts or recesses on several sides, or possibly must have a cored hole. Such removable inserts in a cavity should be provided with suitable ejection means and should have provisions for cleaning on all sides, in view of the penetrating ability of some molding compounds.

The *construction material* for the mold cavity should best be hardened tool steel, highly polished, and chrome plated. Hardening should generally be to 56–58 Rockwell C, depending on the fragility of the cavity section. Cavities may be hobbed, machined, or formed by spark erosion techniques, depending on the configuration of the part. Inasmuch as hobbed cavities can only be case hardened, the surface of such cavities can be dimpled or distorted by the pressures in the molding cycle. Toolmakers generally recommend machined cavities or cavities formed by spark erosion techniques because they can be made in through-hardened steel and are capable of greater precision. The use of through-hardened steel generally prolongs the life of a cavity, provides maximum rework capability, and eliminates the problem of warpage from heat treating. Chrome plating should be "flash" chroming, to a thickness of perhaps 0.0003–0.0005 in.

For further discussion of the details of mold cavities, see MOLDS.

Details in Other Areas. The die base should be rugged, so as not to deflect during mold closing or introduction of molding compound. Land areas, the surfaces of the top and bottom half of the die which bear against one another, should be particularly well polished and close fitting to prevent escape of molding compound during the time when the compound is liquid and the internal pressures are high. The die base should include leader pins to ensure correct alignment of top and bottom half during each closing. In the case of production molds, the die base should have provisions for die clamps for anchoring to the machine platens. Overall mass should be sufficient to assure reasonable heat retention and a uniform temperature profile.

In transfer molding, the transfer pot and plunger, generally cylindrical, should have a diametrical clearance of approximately 0.001 in./in. of diameter. With one or two small half-round O-ring type grooves circumferentially about the transfer plunger, considerable plunger wear can be tolerated without leakage of molding compound, if it is allowed to accumulate and remain in these grooves during any production run. The pot should be hardened, polished, and plated. The plunger should be well polished and hardened, but not plated. For relatively small cavities runners can be small and somewhat flat to encourage maximum heat transfer to the molding compound as it flows down the runner and through the gate, in order to minimize cure time in the cavities. For larger devices, however, runners should be designed with maximum cross-sectional area for minimum cross-sectional circumference, in order that the material not receive too much heat as it goes through the runners and gates into the cavity; in this way the material fills the cavity long before it cures. Runner sections are often provided with ejector pins to ensure that the molded parts as well as the cull and runner will be readily removed in one piece.

In production molds heating may be by steam, hot oil, or electricity. By far the most popular today is electric-cartridge heating, which is clean and convenient. Electric heating cartridges must be adequately distributed under and over the cavity and runner sections to ensure a uniform heat profile on all surfaces with which the molding compound will come in contact. Either thermocouple or filled-bulb temperature-sensing instruments are satisfactory. Sensing elements should generally be located fairly close to one or more of the electric heating cartridges to minimize time lag and wide swings of temperature. The total wattage of electric heating for any given mold generally should be such as to raise the temperature of the mold from ambient to about 300°F in less than 1 hr. Over-temperature cutouts are recommended to eliminate any possibility of mold overheating from a runaway controller or a faulty electrical circuit.

Heating with circulating oil or with steam calls for adequate channeling of the areas beneath and above cavities and runner sections. Some molding compounds and molded parts require chilling before ejection. Steam heating plus water chilling, or circulating hot-oil heat with provisions for cooling the oil, are then required. In connection with the heating provisions, it is well to remember that molds must have sufficient metal to give good heat stability despite variations in ambient temperatures during cycles.

Break-In Procedures for Molds. A critical part in the life of any mold is the first few cycles through which it is used. The mold should be clean of shipping oil, moisture, plating acids, etc, before it is placed in the press. Following mounting in the press the mold should be brought to about 300°F or the expected molding temperature, and the temperature repeatedly checked with a surface pyrometer at various points to ensure that no localized overheating is developing. Some smoking from a new mold may be observed as the last vestiges of oil or other contaminants volatilize from the surface. At about 300°F, amounts of unrefined carnauba wax or other recommended high-temperature release agents should be applied to all surfaces of mold base and cavity, pot and plunger, with which molding compounds will eventually come into contact, intentionally or unintentionally. Excess release agents should be blown off with an air gun, and a shot should be molded using either the molding compound for the part to be made, or a good general-purpose phenolic resin. At least twice the anticipated cure time should be allowed to minimize possibilities of opening the mold when the material is only partially cured. The molded part should be examined to see if cavities are filling adequately, whether flash is appearing where it should not, whether ejector pins are positioned correctly, etc. Such inspection may indicate a need for additional venting, additional die supports, further machining of cavity sections, greater polishing of gates, etc. Several shots will be necessary to determine whether the mold is ready for a production run or whether it needs additional tooling.

If further machining of cavity or runner sections is necessary, it is well to repeat the application of release agent, particularly on any new exposed steel, prior to molding. Too much waxing can lead to further stickage. Following the waxing at break in, normal lubricants and release agents incorporated in the molding compounds will suffice to assure ready release of the molded parts.

During these preliminaries it is desirable to close the mold under pressure, loosen all the mold clamps, and then retighten them, to allow for relief of stresses caused by expansion of various parts of the mold at the operating temperature.

Care and Maintenance of Molds. Probably 90% of tooling problems with molds are from failure of the operator to thoroughly clean the parting surfaces during normal operation. If stickage develops during normal cycling, a light application of release agent at the stickage point will generally remedy the situation. But if stickage persists, the operator should stop further production and the cause of stickage should be determined. It may be from various failures: local collapse of the cavity, causing an undercut; flaking off of chrome plating, requiring polishing and rechroming; damage to that particular section of the mold, perhaps from closing the mold on hard flash or some other hard object.

The operator should visually check the mold before every cycle. Excessive wear on land areas, at gates, or in cavities may be caused by use of a particularly abrasive molding compound. If that particular compound must be used, more frequent replating of the mold is recommended.

The operator should also check that the mold is opening and closing without evidence of misalignment. Visual inspection will readily indicate whether leader pins are rubbing heavily on one side or another, or whether a mold appears to be causing the platen to shift as the mold comes metal to metal. Such indications should be corrected by repositioning the mold in the machine.

Mold storage between production runs should be at room temperature, and with the mold surrounded by plastic to minimize accumulation of dust or other foreign material. When such molds are put back into operation, generally at least one application of wax or release agent is advisable to minimize stickage problems.

Automatic Molding Machines

A natural progression in the art of compression and transfer molding is the development of fully automatic molding machines capable of unattended operation. The basic machines described above still require an operator to deliver the molding compound to the machine, and to remove the molded parts from the mold area following ejection from the cavities. The operator is also needed to actuate the next molding cycle. Automation of these steps has been quite successful in both compression and transfer molding over the past two decades. Such automatic machines, whether for compression molding or for transfer molding, have many features in common which are necessary additions to the basic equipment needed for manual or semiautomatic molding.

Recycling. Automatic machines are designed to actuate the next molding cycle upon satisfactory completion of the preceding cycle. In automatic compression molding, when the parts have been pushed from the cavities by ejector pins, stripper combs and/or air blasts are generally used to collect the molded parts from the ejector pins and remove them from the mold area, generally dropping the parts into bins or baskets. Usually a mechanical or electronic safety device is utilized to determine that no molded parts remain in the mold area. This device generally sends the electric signal to start the machine on its next cycle. Should a part remain in the mold area, the safety device will not indicate an "all clear" condition and the machine will shut itself down.

The next step to be automated is feeding the charge of material into the cavities. For automatic compression machines, feed mechanisms are provided which generally meter a charge of compound from a hopper, and deliver such metered charges to each of the cavities in the open mold. The delivery is often by means of a sliding plate containing a number of cups, the cups being placed on the plate so that they will discharge the material from the bottom of the cup when the plate has reached a position directly over the mold. The plate is then retracted, and when it is clear a limit switch is tripped to allow the automatic machine to take the next step in the molding cycle, then close.

In automatic transfer molding, especially on smaller machines with capacities to approximately 60 g, material feed is generally effected from a hopper or preform bin to the transfer pot. Because only one charge of plastic must be delivered, the feed mechanisms on automatic transfer-molding machines are generally simpler than those on automatic compression-molding machines which must be capable of accommodating varying numbers of cavities and varying sizes of capacities. On larger automatic transfer-molding machines, with capacities from 50 to 300 or more grams, the material must be preheated between the hopper and the pot, usually with high-frequency methods. On such machines the feed mechanisms become more complex and must be very rapid to obtain maximum benefits from the process.

Fig. 15. Stokes model 741 automatic transfer-molding machine with infrared preheating.

After the charge has been delivered and the delivery mechanism is out of the way, the machine can go through the actual molding cycle, of slow-close, breathe, dwell, cure, and open, just like any semiautomatic machine.

In fully automatic transfer molding, where runners are utilized, automatic degating is possible; parts and runners are positively ejected in different planes. One tray carries away the molded parts, dropping them into a suitable bin, while another automatic tray carries away the runner and cull system and drops it into a waste bin.

Most of the larger automatic machines for compression and transfer molding use auxiliary preheating. Electronic preheating or infrared preheating attachments are common in automatic compression-molding machines of 50-ton size or over. These attachments shorten the cure time, and make a single installation more productive. Figure 15 shows the Stokes model 741 fully automatic compression-molding press with infrared preheating.

Safety. Manual or semiautomatic machines have an operator who can generally oversee the molding operation, and can quickly take any necessary steps to minimize possibility of damage to mold, machine, or operator. The operator can blow off all flash from the open mold between each cycle, and can take as long as necessary to ensure that the mold surfaces are clean. He can be sure that the charge of material going to the cavities or to the transfer pot is correct, and that the charge does not contain any impurities which might cause harm to the machine. In automatic operation, on the other hand, the machine may be monitored by an operator only once in several dozen cycles, as the operator makes his rounds among many machines. It is therefore vital that the machine have built-in safety interlocks to ensure that no step of the molding cycle proceed without the successful completion of the previous step.

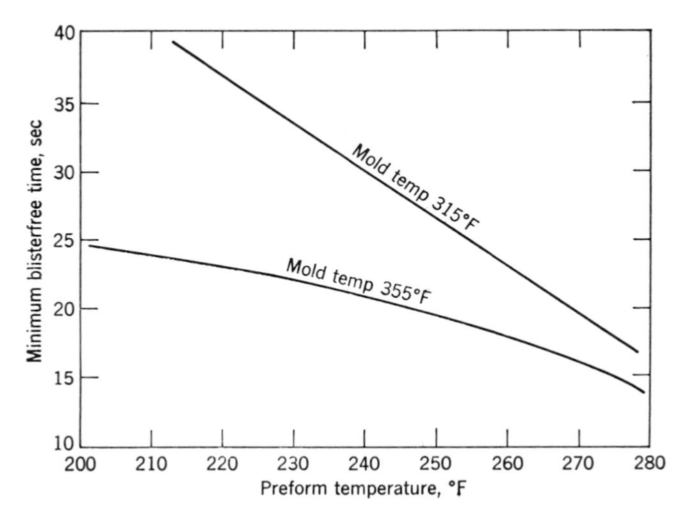

Fig. 16. Minimum curing times using preforms preheated electronically to various temperatures, for the mold temperatures indicated. Plunger-molded caps are made from a general-purpose phenolic resin in this example.

Safety features may be built directly into the mold. In compression molding, where a certain amount of flash on each molded part is inevitable, the design of the land areas and the overflow vents must be such that the flash will remain with the molded part, and not with the mold. Additionally, in automatic molding, both top and bottom ejector pins are recommended to ensure that the part will be fully ejected where possible. The part should be so designed that the need for side draws is eliminated, to avoid too many complexities in the mold itself.

Automatic Transfer or Thermosetting Injection. For many years, the highest production rates in the molding of thermosetting plastics have been achieved through the use of fully automatic compression-molding presses with either dielectric or infrared systems for preheating the compound. Almost equally fast production speeds could be reached in semiautomatic plunger molding with dielectric preform preheating. However, both these techniques have a major limitation; the time interval between the end of preheating and the instant the material is actually molded in the cavities. It is a simple matter to show the effect which this single yet important limitation has on each of the methods.

For example, in automatic compression molding, in which preheating takes place outside the mold area, as long as 15 sec elapses while the hot powder or preform is transported to and placed in the mold, the transport device retracted, and the press closed. Similarly, in high-speed semiautomatic plunger molding, the time interval from preheater to mold cavities, during which the hot material is both cooling and/or curing, is often 10 sec or longer. If the temperature to which a typical thermosetting compound is to be preheated is 200°F, the material will remain fluid enough to mold for perhaps as much as 30 sec. It will then tend to set up regardless of the additional heat or pressure applied in the mold. If the same material is preheated to 250°F, the time available to mold the material before it cures may drop to 15 sec; at 300°F, the time available to mold it may drop to 5 sec.

At the higher temperatures, and with the corresponding shorter time available for molding, curing times are also shorter. Additionally, with higher preheating temperatures, mold temperatures can be increased with less danger of burning the molded part from too great a temperature differential between the mold and the incoming plastic. The data shown in Figure 16 illustrate the shorter blisterfree cure times possible with higher mold temperatures, or with higher preheat temperatures, or with both, when molding a typical phenolic molding compound.

Recognition of this characteristic behavior of thermosetting compounds, and of the essential requirement of getting higher and more uniform preheating of the molding compound before introducing it into the mold, then rapidly getting the pre-

Fig. 17. Stokes model 751 Injectoset press for automatic transfer molding of thermosetting compounds.

heated material into the cavities, has led to development of several new and unusual processing techniques.

Several special techniques worthy of mention are discussed below. It is interesting to note that these methods are departures from automatic compression-molding techniques, and have relied on closed-mold techniques. Injecting material into a clamped mold cuts down the time interval between preheating and molding, thereby permitting higher preheating, yielding shorter cures, and allowing faster cycles. Whether the new closed-mold techniques are called transfer, plunger, or injection molding of thermosets is purely academic, inasmuch as they all utilize preheating and rapid, automatic transfer of the plastic into the cavities.

Fig. 18. Ten-ton horizontal Lewis injection-molding machine.

Injection Transfer or Screw-Plunger. The Stokes Injectoset automatic molding machine, introduced in 1963, is a vertical-clamp downward-acting machine with an upward-moving transfer plunger (Fig. 17). A horizontal extruder-type screw introduces a metered charge of preheated and plasticated material into a transfer pot in each cycle. Uniform heating of the plastic is accomplished within the extruder barrel or the chamber in which the screw rotates, both through the buildup of frictional heat due to mechanical shear of the material and through the external heat input from electrical-resistance band heaters. Material is introduced into a gravity-fed hopper in granular or other loose form rather than in discrete compressed preforms. The screw rotation advances the plastic, and, between cycles, drives plasticated compound against the nozzle end until the screw is forced back away from the nozzle to a preset distance.

At the time for delivering the charge of plasticized material into the transfer pot, the screw is rapidly advanced to the nozzle end, the distance preset for the desired shot size. As soon as the plasticized material has been charged into the transfer pot, the plunger moves upward to drive the material through runners into the cavities, while the screw prepares additional material for the next cycle.

Other features of the machine include means to retract the entire screw chamber and to expel the accumulated plastic, should it harden in the chamber, thus plugging the machine. Molds may utilize automatic degating techniques. Conventional automatic compression press stripper comb systems are used to remove molded parts, culls, and runners from the ejector pins in the open mold. The machine cycles fully automatically and is designed for essentially unattended operation.

Fig. 19. Cold-plunger Lewis automatic injection-molding machine (50-ton capacity).

Temperatures of the plasticated material in the barrel at nozzle end may run as high as 240°F or slightly higher, depending on the specific molding compounds. Inasmuch as an inventory of material larger than that required for a single shot is maintained in the plasticating chamber, temperatures in that chamber must be kept below the point at which the material will progress to full cure exothermically in the chamber before being injected into the mold.

The Injectoset is available with clamp-system capacities from 50 to 250 tons and with shot size capacity from 110 to 660 grams. A wide range of molding compounds has been successfully run in such presses, including glass-filled diallyl phthalate, urea–formaldehyde, and rag-filled phenol–formaldehyde resins. Parts with sections up to 1 in. thick can be cured in 50 sec. Transfer pressures are 10–20% lower than those required in conventional semiautomatic transfer-molding equipment because of the more thoroughly plasticated resin produced.

In 1966 such machines were offered not only by Stokes Division of Pennsalt Corp., but also by Rodgers Hydraulic Press Manufacturing Co. (the Rodgers-Egan machine) and by several overseas companies.

Ram Injection. Also introduced in 1963, the Lewis automatic injection-molding press for thermosetting plastics (Fig. 18) utilizes a preplasticating injection chamber or cylinder, jacketed for circulation of water or other heat-transfer medium to maintain cylinder temperatures in the range of 175–250°F. Using a ram-injection technique common to many of the older thermoplastic injection machines, the Lewis machine injects the plasticated compound directly into the runner system of the mold, and thereby eliminates the separate transfer pot and plunger found in the screw-plunger

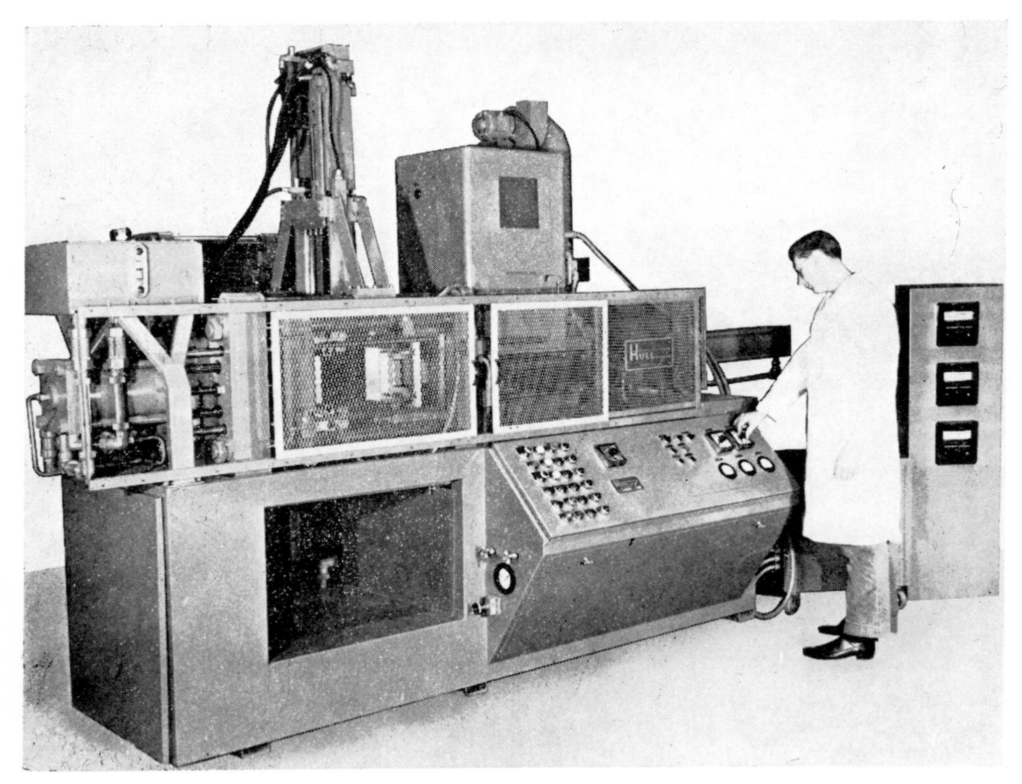

Fig. 20. Hull 99C automatic transfer-molding press with electronic preheater.

machines described above. The Lewis machines are available in both vertical and horizontal configuration, in sizes with mold-clamp capacities from 28 to 200 tons. Cycle times are one-third to one-ninth those of conventional semiautomatic equipment even with difficult compounds such as flock-filled phenolic and glass-filled diallyl phthalate resins. Other materials which have been molded include silicone, epoxy, and urea–formaldehyde molding compounds.

Another new development in this field is a 50-ton cold-plunger injection machine for high-speed, low-temperature molding of glass-reinforced alkyd resins (Fig. 19). The machine was developed by The Lewis Welding and Engineering Corp. in conjunction with Allied Chemical Corp. It is basically a four-station rotary molding machine. It preheats material to temperatures in the range of 100 to 165°F, and can achieve significant increases in production rate of small molded parts. Since the machine is designed specifically for molding alkyd resins, which are extremely sensitive to temperature during processing steps, the use of higher preheat temperatures is impractical.

El-Tronics, Inc., of Warren, Pennsylvania, also offers a multistation rotary ram-injection machine which operates on much the same principle.

A refinement of the screw-plunger process is the in-line screw concept for thermosetting plastics, introduced in 1965 by New Britain Anchorwerke, and subsequently offered by Stokes and several foreign manufacturers. In this concept, the material is plasticated by a reciprocating screw in much the same manner as in the screw-plunger system described above. For each cycle, a metered amount of the plasticated material is forced from the barrel by the screw directly into the runner system of the mold, in a manner very similar to that used in injection molding with thermoplastic materials. Although pressures of 17,000–20,000 psi on the material are often used, as compared to 6,000–12,000 psi with the screw-plunger method, the in-line approach offers faster cycling. Controls become very critical; they minimize the possibilities of material setup in the barrel or nozzle.

Automated Plunger Transfer. The Hull 99C automatic transfer-molding press was also introduced in 1964. The 99C (Fig. 20), unlike the smaller 99A and 99B models, uses a specially constructed and integrated LaRose electronic preheater to raise the temperature of molding charge prior to transfer. Equipped with a horizontally acting injection or plunger transfer system, the 99C is available either with an integral preformer accepting granular material, hopper-fed, or with a Syntron bowl feeder accepting cylindrical preforms. Preforms are fed onto a pair of rotating rods located directly above an opening in the horizontal transfer tube or pot. These rods comprise one electrode of the electronic preheater, while a fixed electrode is directly above the preforms. The preheating cycle is synchronized to start during the curing cycle of the previous charge of molding material.

The preforms are heated by 100-mc radiofrequency energy to temperatures around 300°F in 5–9 sec, depending on the material. Inasmuch as only the single charge to be molded is preheated (one each cycle), these unusually high temperatures are possible since there is no inventory of hot material. Mold temperatures can be as high as 400°F. During the preheat period, the previous charge of material completes its cure, the mold opens, parts are ejected, the transfer plunger retracts, and the mold closes. This phase of the cycle is completed at the precise moment when the electronic preheating is completed. The preforms are instantly inserted into the transfer tube or injection cylinder, and the plunger transfers them to the runners and cavities.

With such high preheat temperatures and fast press action, transfer times from completion of preheating to the full cavity are between 2 and 3 sec for shots up to 100 g (3.5 oz) and cure times are as short as 7–10 sec. Overall cycle times for electrical-grade phenol–formaldehyde molding compounds are typically 15 sec, while overall cycles of 20 and 25 sec are typical for melamine–formaldehyde and diallyl phthalate resins in multicavity connector molds with shot sizes up to 100 g. Alkyd, epoxy, and urea–formaldehyde molding compounds have also been run successfully on these automated transfer plunger machines. Pressures on the material are generally from 3000 to 5000 psi, resulting in minimum mold wear and smaller clamping force for a given projected area of parts.

The currently available press is a 75-ton clamp capacity machine utilizing an integral 7.5-kW dielectric preheater operating at 100 mc and capable of handling shot sizes up to approximately 300 g of general-purpose phenol–formaldehyde molding compound on a production basis.

Summary. Each of the machines discussed permits high-speed economical production of thermosetting plastic parts at rates which render practically obsolete fully automatic compression machines and semiautomatic compression and transfer machines. None is designed for automatic insert loading, so there is still a place for the semiautomatic machines. The trend to these new machines, however, appears inexorable.

Special Applications

Considerable ingenuity has always been demonstrated in the application of plastics for a wide variety of end uses. And, while plastics have often been substituted for historically used materials, such substitutions in many instances yield better products more economically manufactured. Recall, for example, the old porcelain-type insulators which formed a part of electrical lamps and appliances around the home. Porcelain proved a good, rigid, high-temperature material and it could be fired and glazed to present a reasonably attractive appearance. Today very few porcelain fixtures are found in the United States, because molded thermosetting plastics have proved stronger, less fragile, less expensive, less bulky, and capable of many esthetic design and color possibilities.

To illustrate the ingenuity of the industry, a few specific special techniques which have broadened the use of thermosetting plastics will be discussed.

Fig. 21. Mold and work-holding fixture for encapsulating 96 capacitors per cycle and capable of encapsulating well over 20,000 devices per shift. Courtesy Hull Corp.

Melamine Dinnerware. Melamine–formaldehyde is a very hard and scratch-resistant material available in a variety of stable colors with heat-distortion point well over 300°F (see also AMINO RESINS). It is a natural material for the manufacture of relatively inexpensive so-called unbreakable dinnerware. In melamine dinnerware the use of decorative foil overlays, introduced during the cure cycle, have required more sophistication in the control systems. In a typical dinnerware molding cycle where overlays are used the preliminary cycle involves the operator placing the molding compound in the cavities, actuating the "preliminary cycle" push buttons causing the press to close, breathe, dwell, and close for a partial cure, after which the press opens automatically. The operator then lays the foil over the partially cured parts, and actuates the secondary-cycle buttons. The press then closes, perhaps at a different rate and with a different period of slow-close, breathe, breathe–dwell, and cure, than in the preliminary cycle. When the press finally opens, the molded parts with the decorative overlay integrally molded on the surface of the part are removed from the machine.

Dinnerware molding, in addition to requiring greater sophistication of controls, has required extreme sophistication in the deflashing equipment. Flash must not only be removed, but the flash line must be thoroughly buffed and polished so that there is no rough surface to retain food particles during washing. Tumbling is rarely used; instead highly automated flash-removing jigs and tools, often custom built, are employed.

Encapsulation of Electronic Components. The availability of soft-flowing epoxy molding compounds has led to another significant trend, the development of special transfer presses for encapsulating by transfer molding. Small electronic components and complete electronic circuits, as well as electrical coils and transformers, can be encapsulated in thermosetting plastic insulating materials very economically in large quantities (see Fig. 21). Because the encapsulating compounds must be soft flowing, encapsulating presses generally have a larger die space for given tonnage, and often have more sensitive pressure controls than do conventional compression- and transfer-molding presses.

Screw Closures. Some years older than the special applications described above, but still used extensively, is the technique of automatically molding screw-threaded caps and other screw closures for use in the cosmetic and beverage industries and other applications requiring threaded plastic parts. Such parts are generally made by compression molding, with the machine and mold specially equipped with unscrewing mechanisms to rotate the force within the cavity during the breakaway or first opening of the mold following cure in such a manner as to unscrew the molded part from the threaded mold core. Additional wiper arms with resilient surfaces move past the parts as soon as they are exposed during ejection to ensure further unscrewing until the parts fall onto a tray below. The parts are then removed from the mold area in a conventional manner, and the next cycle begins. Automatic-closure installations with as many as 144 cavities have operated successfully for more than 25 years. Transfer-molding techniques may also be used in automatic-closure molding but are less common because the runner system must be separated from the closure before the unthreading operation begins, and because the gate scar may be unsightly if the closure is for an esthetic application. Mechanically, however, unscrewing means can be incorporated in the machine and mold for such requirements.

Economics

Approximate values of the equipment which might be considered for a molding operation are given in Table 2. Although such figures are valid for preliminary cost studies, based on selling prices in effect during 1966, it is advisable to request up to date prices from the manufacturers of machinery and tools as the cost study progresses.

Table 2. Approximate Prices[a] of Typical Thermosetting Molding Equipment

Characteristics					Price, $
Molds					
hand, single, small-cavity (diam 1 in.)					400 and up
hand, single, large-cavity (diam 4 in.)					1,000 and up
production, four-cavity (diam 3 in.) part					5,000 and up
production, fifty-cavity (diam 1 in.) part					8,000 and up
Automatic preformers					
10-ton (diam of representative preform 1 in.)					2,500
20-ton (2 in.)					3,500
35-ton (3 in.)					5,000
75-ton (4 in.)					8,000
Electronic preheaters					
1-kW					500
2-kW					1,800
5-kW					3,000
10-kW					5,000
Molding presses[b,c]					
	10–20 tons	25–30 tons	50–75 tons	100–150 tons	200–300 tons
manual compression	1,000	3,000	5,000	9,000	12,000
manual transfer	2,000	4,000	6,500	11,000	14,000
semiautomatic compression	2,500	5,000	8,000	12,000	20,000
semiautomatic transfer	3,500	6,500	10,000	15,000	25,000
fully automatic compression	4,000	6,000	15,000	20,000	25,000
fully automatic transfer (injection)	5,500	8,500	30,000	37,000	45,000

[a] Prices as of 1966.

[b] Molds are either transfer or compression types, of hardened tool steel, highly polished, and chrome plated.

[c] The complexity of the part, the tolerances required, etc, affect the price of the mold. These prices assume relatively simple parts with standard right-angle corners, spherical surfaces, etc.

Cost studies require estimates of material costs and cycle times. The material costs can generally be determined accurately from the volume of the part and the amount of the appropriate material Thermosetting compounds vary from approximately 25¢/lb for general-purpose phenolic resin to several dollars per pound for some of the more exotic silicone, diallyl phthalate, or epoxy resins. Because of the thousands of molding compounds available, suppliers should be consulted for current prices of the required types of formulation.

Knowledge of the molding compound to be used and of the size of the part will then make possible an estimate of the cycle time. An experienced molder should make the estimate, basing it on part size, on the planned molding techniques, and on the extent of auxiliary or automated equipment which will be available. Experience is the most valuable factor in estimating cycle times, as there is no known method of calculating with any accuracy the minimum cure time of a given thermosetting plastic part. Generally the cure time for a single-cavity mold of a given part will be quite comparable to the cure time for a multicavity operation of the same part. Once the cycle time is estimated, it can be related to the required production rate to indicate the number of cavities which will be required in a given production operation. With this knowledge, it is possible to obtain an estimate of the mold, and of the size machine required. At this point, other costs such as floor space, anticipated electric power, water, and air services, plus other shop costs, and the labor (one man for one or two semiautomatic machines, or one man for four to eight automatic machines, for example), will enable the cost study to be completed with reasonable approximations.

Factors for wastage of material are often 5–10% in compression molding, and anywhere from 5 to about 75% in transfer molding, depending on size of part and number of cavities. Machine utilization is generally estimated at 70–80% of capacity for either semiautomatic or fully automatic machines over any length of time. For simple cost estimates, amortization of the machine may be taken from five to ten years. Maintenance of the machine is often estimated at 5% of the initial cost each year. Amortization of the mold is often taken over a two-year period. Although the mold will usually last many years, allowance must be made for possible design changes in the part, or even discontinuance of the particular part. Mold maintenance may be estimated at approximately 5% of the initial cost per year.

All the above factors are fairly straightforward, and the cost estimates are also straightforward. But they fail to take into consideration a wide variety of less tangible factors. Because each molding requirement is unique, it is impossible to include in any general discussion all of the factors which must be considered. Of course, molding has already been compared with other manufacturing processes, such as machining (qv) or stamping parts from laminates, both for ease and accuracy of production and for economics.

There may also be alternative molding approaches to the production of a particular part. For example, in some instances a molded plastic part must have another plastic or metal part attached to it, often by a threaded screw. There are several choices. (a) After molding, including the molding of a straight-sided hole, the hole may be threaded by a tapping operation, mechanically or manually. (b) The part may be produced with a female threaded insert molded in. (c) A threaded insert, possibly with knurled OD, can be pressed into a molded hole. (d) The mold can incorporate a threaded core which can be manually or automatically rotated on completion of the cure cycle such that the threads are effectively molded in the hole. (e) The two objects may be fastened together by some other means, including adhesives or clamps. (f) There is the possibility of molding the two parts as one, thus eliminating the need for any threaded hole. Obviously, each of the above possibilities can be analyzed in terms of the number of parts to be produced, the time needed to perform the threading operation, the cost of the necessary equipment, etc. No one solution will apply to all such requirements. But for any given instance, one solution probably will yield optimum economics.

Productivity. There are times when production requirements increase for an item which is currently being molded. Perhaps it is not possible to make major changes in machinery or molds. There are ways to increase the efficiency or productivity of most cycles. In semiautomatic operations, use of a preformer to eliminate the handling of loose powder often cuts cost. Use of high-frequency preheating shortens overall cycles. A loading fixture may speed up loading of inserts into the mold cavities. Shrink fixtures used after molding may make possible removal of parts after a shorter cure time.

In automatic operation, further evaluation of new materials may find a material which can be cycled more rapidly than the material in use. Re-evaluation of the setting of various process variables, including temperature, pressure, and speeds, may save a few seconds in the overall cycle and therefore bring about an increase in productivity. Opening gates and vents and enlarging runners in transfer molds may enable faster filling of cavities and thus shortening of cure time.

Future Outlook

Because most of today's "advanced" techniques and equipment have evolved from a need to meet various design, production, or economic requirements, it is probable that future developments in equipment and techniques for processing thermosetting plastics will be brought about by today's unsolved needs.

From the point of view of economics, it is desirable to produce plastic parts in shorter cycle times using less expensive machines and molds. Fast cycles will require newer resin systems capable of practically instantaneous cure. At the present efforts are being made to effect curing through the use of laser energy. It is therefore conceivable that at some time in the future material will be enclosed in a cavity and laser energy will trigger an instantaneous crosslinking, with the result that very little cure time in the mold is necessary. As cure times become shorter, faster-acting molding machines will become necessary. Whereas speeds of platen and plunger movement today may approach 500–600 in./min, these speeds must be increased many times as faster-curing materials become available.

Presses today must be heavy in order to exert the necessary pressures. New materials of the future will be moldable at much lower pressures, and equipment therefore will be lighter, faster, and less expensive. Control systems will be solid-state, and machines will be set up using computer-programming techniques with automatic feedback to control the process variables.

As materials require less pressure, molds will be manufactured less expensively. They will not need to be strong enough to withstand the high pressures used today, and they will not need to be hard enough to withstand the erosion of material flowing under high pressures and with high viscosities. Although mold-manufacturing techniques involve a considerable amount of precision metalworking, it is probable that an ideal plastic will be developed with strength approaching that of steel and with heat-conductivity characteristics approaching those of metals. Such a plastic will then be used for manufacturing molds much more economically than steel.

Techniques have already been developed to formulate the basic molding compound by actually blending the resin and catalyst and filler in the transfer pot immediately prior to transfer molding. It may therefore be possible for the chemical formulating and the molding operation to be combined into a single plastics-processing facility, with the additional possibility of varying the types and quantities of chemicals introduced into the pot to effect desired end properties in the molded part.

Bibliography

Modern Plastics Encyclopedia, McGraw-Hill Book Co., New York, Vols. 42–45, 1964–1967.

Plastics Engineering Handbook, Reinhold Publishing Corp., New York, 3rd ed., 1960.

R. W. Bainbridge, "Physical Characteristics of Compounds, Compression vs. Transfer and Injection Transfer Method," *SPE Tech. Papers* **12,** VIII-2 (1966).

E. F. Borro, Sr., "Compression and Transfer Molding," *SPE Tech. Papers* **8,** 19–2 (1962).

J. M. Brach, "When You Plan to Go Automatic," *Paper, Soc. Plastics Eng., Reg. Tech. Conf., Chicago, Dec. 12–13, 1966.*

D. A. Daniels, "Recent Developments and Applications in High-Frequency Preheating of Thermosets," *Paper, Soc. Plastics Eng., Reg. Tech. Conf., Chicago, Dec.* 12–13, 1966.

F. Florentine et al., "Effect of Certain Variables on Preheating of Phenolic Molding," *SPE Tech. Papers* **10,** XIV-3 (1964).

H. M. Gardner et al., "The Injection Molding of Thermoset Plastics," *SPE Tech. Papers* **12** (1966).

D. F. Hoffman, "Injection Molding of Thermosets," *Paper, Soc. Plastics Eng., Reg. Tech. Conf., Chicago, Dec. 12–13, 1966.*

J. L. Hull, "The Case for Electronic Preheating in Automated Transfer Molding," *Mod. Plastics* **44** (9), 121 (May 1967).

J. L. Hull, "Compression and Transfer Molding," *Plastics Design Process.* (June 1961).

J. L. Hull, "A New Concept—Molding Small Parts," *SPE J.* **13** (5), (May 1957).

J. L. Hull, "New Developments in Automatic Transfer Molding of Thermosetting Compounds," *SPE Tech. Papers* **8,** 19–3 (1962).

A. W. Hunt, "Hot Preheat, the Key to Ultra-High-Speed Thermosetting Molding," *SPE Tech. Papers* **12** (1966).

A. W. Hunt, "Vacuum Transfer Encapsulation," *Mod. Plastics* (March 1966).

A. L. Maiocco, "Transfer Molding, Past, Present, and Future," *SPE Tech. Papers* **10,** XIV-4 (1964).

Y. Morita, "Screw Injection Moulding of Thermoset," *SPE Tech. Papers* **12,** XIV-5 (1966).

H. R. Simonds, ed., *Encyclopedia of Plastics Equipment*, Reinhold Publishing Corp., New York.

S. E. Tinkham, "New Ideas in Molds for Thermosets," *Mod. Plastics* **44** (7), (March 1967).

E. W. Vaill, "Injection and Extrusion of Thermosets," *SPE Tech. Papers* **12,** VIII-6 (1966).

B. P. Wenninger, "Tooling for New Process Techniques—Thermosets," *Paper, Soc. Plastics Eng., Reg. Tech. Conf., Chicago, Dec. 12–13, 1966.*

R. F. Zecher, "Transfer Molding of New Thermoset Compounds," *SPE J.* **20** (1), 1023 (Jan. 1964).

R. F. Zecher, "What Does Transfer Molding Offer Today?" *Insulation* **13** (1), 37 (Jan. 1967).

John L. Hull
Hull Corporation

INJECTION MOLDING

Injection molding is considered to have begun about 100 years ago, close to the time when Hyatt successfully produced Celluloid; however, it is only within the last forty years that its full potential has been realized. Inherent advantages of the injection-molding process are its high degree of reproducibility and its ability to produce a wide range of products economically (1–9). Since the process was first used, growth in equipment and techniques have paralleled the development of new plastics, rubbers, and resins which can be processed by the injection method.

The Molding Cycle

Injection molding is an intermittent, cyclic process in which particles of plastic material are heated until they become molten. The melt is then forced into a closed mold where it solidifies to form a part shaped in the reverse image of the mold (see also Molds). The manner in which a material resolidifies is one criterion for differentiating one injection-molding process from another. For example, thermoplastics resolidify by cooling, whereas solidification by the addition of heat to the mold implies vulcanization, as in rubber curing, or polymerization, as in the curing of thermosetting resins. Highly specialized machines are designed to be most suitable for the various processes (10–15).

This article will be concerned primarily with the processes peculiar to the injection molding of thermoplastic and thermosetting resins. Transfer molding, which is often considered to be an older form of injection molding, will be compared with the "direct" injection-molding process for molding thermosetting resins. Machines used for injection blow molding are other excellent examples of the adaptation of the ram or reciprocating screw of injection molding. (See also the sections of this article on Blow Molding and on Compression and Transfer Molding).

Multipurpose or so-called universal injection-molding machines, including injection blow-molding machines, are currently following a new design trend by adopting modular design, which has been successfully used in designing other complex automatic machinery. Modular design implies the use of a group of independent machine elements mounted on a common base to form a very specific type of automatic machine. By interchanging certain of these machine elements with others it becomes feasible to produce a new type of machine which can be used for producing other types of end products.

Machinery

The most widely used contemporary machines include: (a) the ram injection-molding machine comprised of a cylinder, spreader, and plunger which forces the melt into the mold (Fig. 1); (b) the plunger- or screw-type plasticator with second-stage injection, comprised of a plasticator, a directional valve, a cylinder without a spreader, and a ram or plunger which forces the melt into the mold (Figs. 2 and 3); and (c) the reciprocating-screw injection machine comprised of a barrel or cylinder, and a screw which rotates to melt and mix the material and then moves forward to force

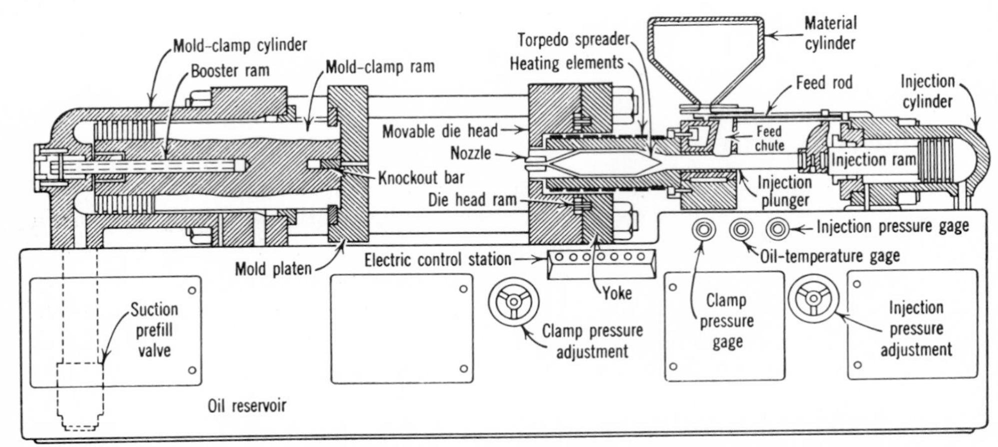

Fig. 1. Schematic view of plunger-type injection-molding machine. Courtesy Hydraulic Press Manufacturing Co.

the melt into the mold (Fig. 4). Other machines found in modern production shops include modifications or combinations of these three basic types. Notable among these modifications are two-stage injection machines in which a stationary screw acts as a feeder and plasticator causing the melt to work against the restriction of a directional control valve. The position of the valve either allows the material to accumulate in the flights of the screw or directs the melt into the injection cylinder. Once it is in the injection cylinder, the charge assists in moving the ram back to its starting position for the injection stroke. A modification of the ram injection machine utilizes a rotating spreader (now largely obsolete) to facilitate the melting of the granules (Fig. 5).

In older machine designs, the plasticator, at first incorrectly called a plasticizer or preplasticizer, was successfully utilized to increase the charge or machine capacity without increasing either the number of cylinders or the cylinder diameter, or the power required to inject the charge into the mold. (The terms "plasticize," "plastify," and "plasticate" as used in molding describe the action which takes place in a cylinder of a molding machine which changes solid thermoplastic particles to a homogeneous viscous plastic melt.) Screw injection has made the older ram unit obsolete in most cases, and, since its introduction, has had a great influence in furthering the growth of the injection-molding process (16–21).

All injection-molding machines are equipped with a pair of mold clamping platens. One platen, the movable end, is moved by either hydraulic, pneumatic, or mechanical means to the clamping position. Each press, contingent upon its degree of sophistica-

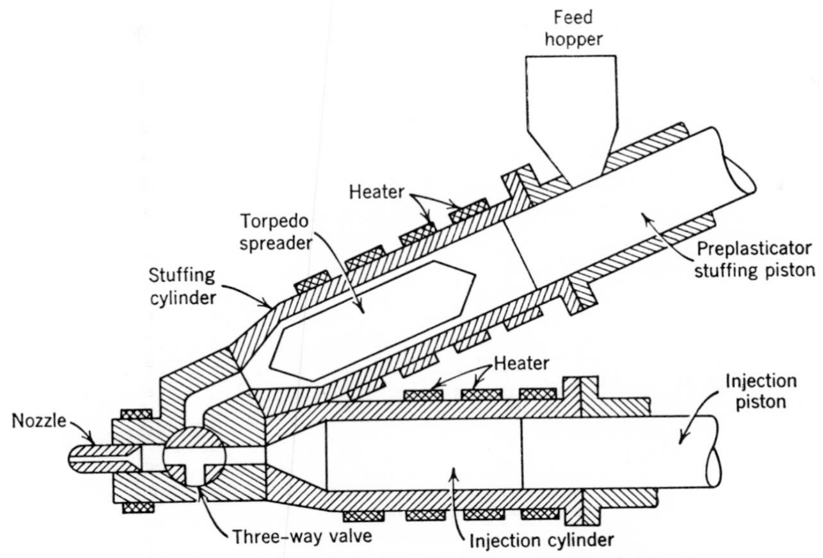

Fig. 2. Schematic diagram of plunger-plunger preplastication system.

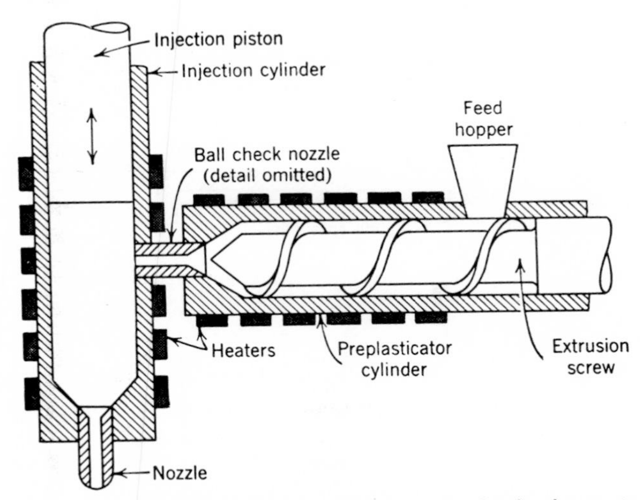

Fig. 3. Schematic diagram of screw-plunger preplastication system.

tion, is also equipped with a number of valves, timers, heating controls, sequencing devices, and safety interlocks for semiautomatic or fully automatic molding.

Molding of Thermoplastic Materials

Thermoplastic materials are characterized by their ability to soften when heated and to return to their original solid state when cooled (22–39). These materials are normally produced in pellet form, but powders are also available. Powders, which were originally introduced to the extrusion process (see also MELT EXTRUSION), can also be utilized in screw injection. Pellets, on the other hand, differ in shape and size according to manufacturers' standards and material type. Pellet configuration for injection molding must (a) be free-flowing, without tendency to bridge or agglomerate in the throat of the machine and (b) be free of gels and other defects such as variations in bulk density due to inhomogeneities in molecular weight. The latter is particularly important when volumetric feeders are used and can also be bothersome when weigh feeders are used on ram-type machines.

The injection-molding machine arrangement most widely used in thermoplastic molding is the horizontal machine in which the clamping and injection units operate

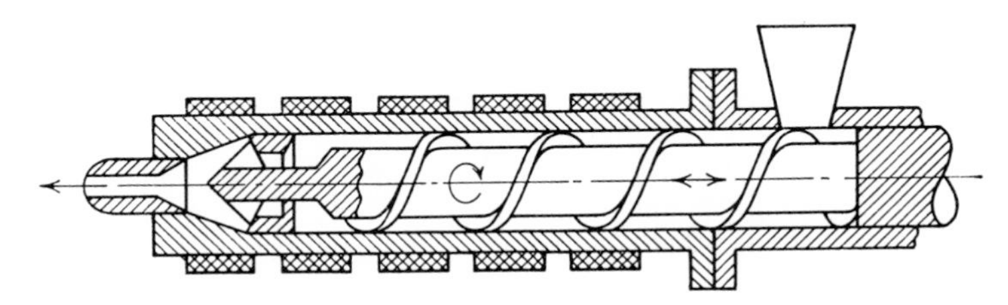

Fig. 4. Schematic diagram reciprocating of screw plastication system.

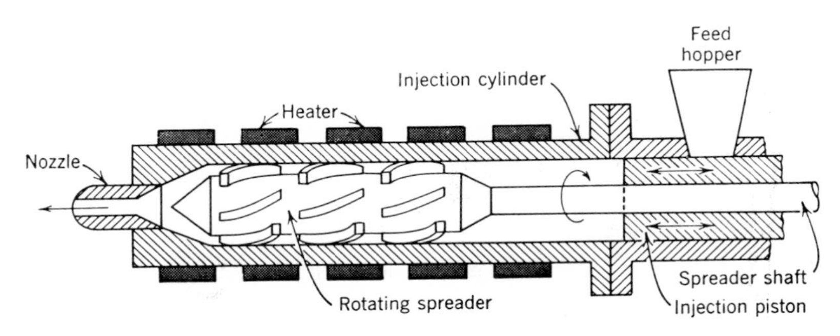

Fig. 5. Schematic diagram of plunger–rotating spreader plastication system.

in the same horizontal plane (40–44). Vertical machines which are designed for specialty operation in the record-making industry are called "boomers" when they exist without clamping units. The axial arrangement (vertical) is also retained in machines with clamping units, particularly where insert loading means are employed. Other types of machines, less widely used, include equipment in which the injection unit is at right angles to the axis of the clamp. In the Lester machines (Fig. 6), material is fed to the nozzle, which is axial to the clamp, from a special plasticating cylinder. Other designs, which preceded this type, allow injection to be carried out at right angles to the center line of the clamp and on the split line of the mold (Fig. 7). Such machines have also been designed with multiple plasticating cylinders, but are few in number because they are so very highly specialized.

Auxiliary Procedures. Drying, blending, and regrinding will be discussed briefly.

Drying. Thermoplastic materials are generally received in a form ready for immediate use. However, hygroscopic materials must be dried prior to being molded in order to reduce the moisture content of the material to a point where it will not affect the molding process or produce defective molded parts. Moisture is either absorbed into the material or adsorbed onto its surface. Cellulosic plastics, nylons,

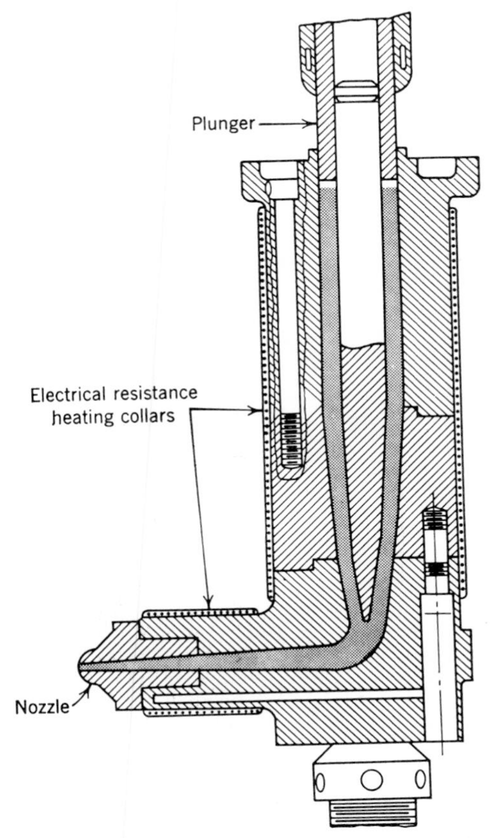

Fig. 6. Lester vertical injection cylinder. Courtesy Lester-Phoenix Co., Inc.

polycarbonates, and acrylic resins readily absorb moisture from humid air. Thus, when these materials are molded without drying, specks, bubbles, blisters, and internal voids result from steam entrapment. Condensate on the surface of pellets is easier to remove than absorbed moisture. The principles and practice of moisture removal are discussed in detail in the article DRYING.

Blending. Another important premolding operation is blending. Blending operations cover a broad spectrum of mixing, masterbatching, and compounding, which includes the operation of color blending (45). These methods are described in greater detail in other articles; see, for example COMPOUNDING; RUBBER COMPOUNDING AND PROCESSING. They will therefore be discussed in very limited fashion here only as they affect the injection-molding process (46).

In dry blending, color pigment (specially prepared for adhesion to the polymer surfaces) is added and mixed with the clear material before it is placed in the hopper of the injection-molding machine. Equipment for dry blending of molding materials is relatively simple. End-over-end drum tumblers, charged with material and slowly rotated (20–30 rpm for 15–30 min) are generally found to be economical. Separate drums for each color are recommended because barrel cleaning is uneconomical. Other equipment includes conical blenders, concrete mixers, and ribbon blenders.

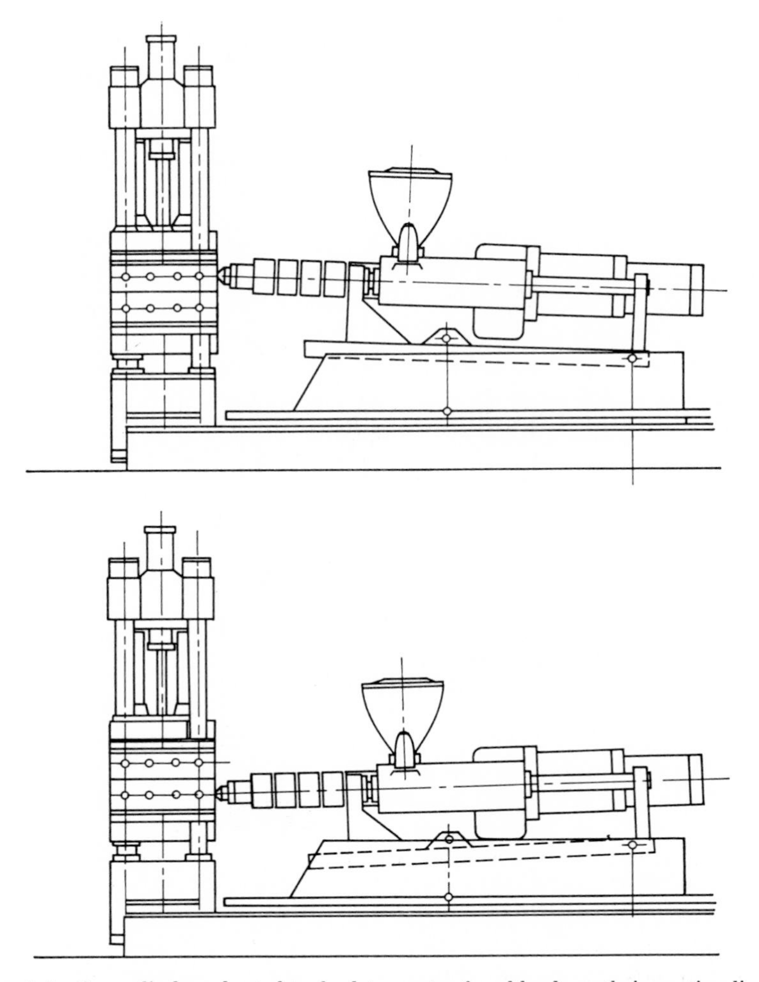

Fig. 7. Screw injection cylinder adapted to feed two sets of molds along their parting lines. Molds are vertically clamped in a compression-type press.

Regrinding. A customary procedure in molding of thermoplastics is the reclaiming of scrap materials from sprues, runners, and rejects. Reground material may be reintroduced into the molding process by incorporation into new, virgin material entering the machine hopper. Certain percentages of uncontaminated regrind of many thermoplastic resins may be mixed with fresh material to improve its flow characteristic without impairing other properties. The reclamation and reuse of scrap material are discussed in detail in the section on Auxiliary Procedures in MOLDING.

Drying, color blending, and regrinding are, with respect to injection molding, principally used in processing thermoplastic materials. Therefore, they have been listed here under thermoplastic operations. Other auxiliary operations, such as feeding methods, and equipment common to injection molding, of thermosetting as well as thermoplastic materials will be described under Molding of Thermosetting Materials. Drives, hydraulic systems, and controls common to all injection-molding machines will also be described in this section.

Characteristics of Thermoplastic Resins. Thermoplastic resins, because of their comparatively high melt viscosities, require relatively high pressures to force molten polymer into the mold; these resins also usually melt over a fairly wide temperature range.

Since thermoplastic resins are notoriously poor heat conductors, it is often desirable to chill the molds in order to speed the solidification process; this constitutes the cooling cycle of the machine. However, a highly chilled mold is never recommended and will almost always be detrimental. Overchilling a mold can result in rapid surface hardening of a thick part while its center is yet fluid. Defects such as sinks, voids, and distortion become inevitable under such conditions (47). Moisture condensing from surrounding air on a very cold mold surface can cause many and varied surface markings on the molded part.

Thermoplastic melts, again in contrast to molten metals, exhibit wide deviations from Newtonian viscosity, and a given pressure difference results in a higher rate of flow than would be experienced with a Newtonian fluid. This flow behavior of polymer melts lead to their characteristic behavior in mold filling. The melt enters the mold cavity as a mass which enlarges outwardly to contact the cavity walls, where it solidifies. The core of the material, however, remains as an elastic fluid and permits more material to enter the mold. Under these conditions it can readily be seen that it is possible to "pack" a mold with plastic when excessive injection pressure is used in combination with a volume of material in excess of that required for the molding. Overfilling, or stuffing, a mold is one cause of "frozen-in" strains in molded parts. Such parts may warp in the mold (47) or break when being ejected from the mold. On the other hand, keeping a packed mold under pressure while the part solidifies can distort or damage the mold. Between the two extremes are parts which warp after removal from the mold, others with dimensions greater than specified because of postmolding expansion, and yet others exhibiting heavy flash at parting lines.

Molding of Thermosetting Materials and Elastomers

Prior to 1940, little work was done to adapt injection-molding techniques to the molding of thermosetting resins. Compression- and transfer-molding operations appeared to be amply able to handle production and there was little incentive to experiment with the phenolic and the urea– and melamine–formaldehyde resins then available. Since that time, however, progress has been rapid. Three distinct methods of injection-molding thermosetting materials were developed in the years prior to 1950 and two others have been introduced since 1964. Jet molding, straight injection molding, and offset molding were introduced earlier; screw-injection and screw-transfer molding are the more recent additions (48–57).

Jet Molding. Jet molding is a patented process (57) consisting of the application of a special nozzle to any standard injection-molding machine (Fig. 8). Although the process is no longer in wide commercial use, a description of it is of historical interest.

In jet molding, the nozzle is heated and cooled very rapidly during each cycle. Two electrodes spaced along the length of the nozzle heat the metal between them to a very high temperature by electrical resistance. This heat is transferred to the resin as it passes through the orifice of the nozzle into a heated mold. The cylinder of a standard plunger machine, with its torpedo or spreader removed, acts as a preheat chamber set at 200°F. Cooling of the nozzle is controlled by water which flows continuously

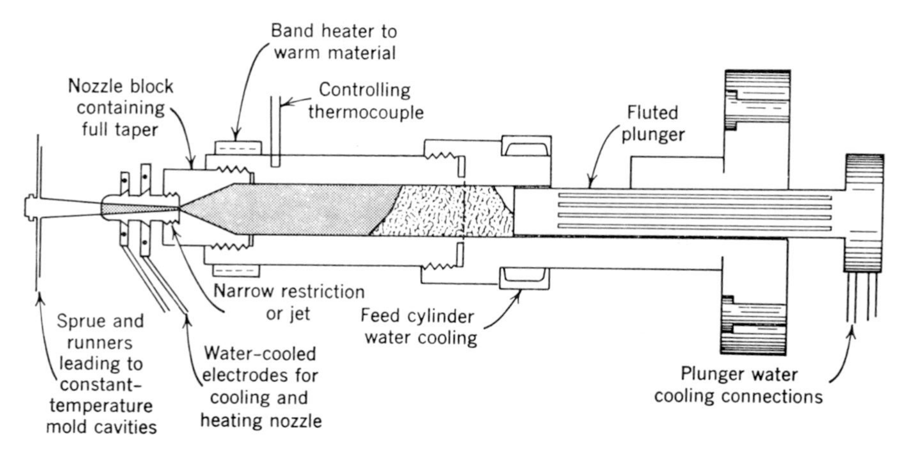

Fig. 8. Diagram of jet-molding cylinder. Courtesy *Modern Plastics.*

through the electrodes and lowers the temperature as soon as the electric current is off, thus preventing polymerization of the material remaining in the nozzle after the injection stroke.

The jet mold is similar to the injection mold for thermoplastic resins but is heated to 300–350°F, whereas the latter is generally cooled. Heat is normally supplied by steam or electric resistance heaters to the mold.

Advantages offered by jet molding of thermosetting resins are fast cure rates, uniform density of molded parts, and the ability to encapsulate or utilize delicate inserts in molding.

Straight Injection Molding. The method for straight injection molding of thermosetting polymers, ie, molding in a straight line, as opposed to offset injection, originated with standard plunger-type injection-molding machines, which were altered to accommodate the thermosetting character of the resins. Jacketed cylinders with streamlined bores, smoothly tapering through to the orifices in the nozzles, were incorporated. These were made without spreaders or torpedos to avoid any tendency that they might have to separate or cause the thermosetting compounds to scorch. The jackets allow either steam, water, or oil to be used as a heating medium for the lower temperatures required by B-stage thermosetting resins. Some designs also used electric resistance heating bands in conjunction with water-cooled jackets for fine temperature control.

Nozzles for injection molding of thermosetting plastics also require special design (55). Whereas a nozzle for thermoplastic resins is tapered from a small opening at its tip to a larger opening at the base of the barrel, a nozzle for thermosetting resins is oppositely tapered. This design gives some assurance that the division between the hot zone (mold) and the cold zone (barrel) will occur somewhere within the confines of the nozzle and that any material which cures in the nozzle can be readily removed by the sprue before the start of a new cycle.

Injection molds for thermosetting resins are operated at slightly higher temperatures than molds used in jet molding. Heat for curing the resin is obtained almost entirely from the mold, with only small amounts contributed by friction of flow through gates or by the preheating in the heating cylinder. In preheating the material, temperatures must be closely regulated to prevent premature curing of the material before it enters the mold.

Offset Injection Molding. Offset injection molding, unlike the other injection methods, is a transfer-molding operation carried out on an altered bed of a horizontal injection-molding machine. The transfer pot, dielectric preform heater, and stationary mold half are mounted to the stationary platen of the machine. The transfer plunger is then coupled to the injection ram or plunger while the movable mold half is mounted to the movable platen in normal fashion. In this modification, the heating cylinder is either removed or left open to facilitate the coupling of the two plungers. Offset molding is so named because of the manner in which material is forced sideways through runners into cavities offset along the flow path.

In operation, preforms are dielectrically heated in a yoke which carries them into position in front of the plunger (see also DIELECTRIC HEATING). As the plunger moves forward, it forces the softened material into the heated transfer pot, and from there into runners and cavities where the material cures rapidly. After the mold opens, the plunger continues to move forward to expel the cull before returning to reload. Some arrangements also provide a heated plunger to prevent chilling the material during the injection stroke.

Recent automatic transfer injection-molding processes have almost totally displaced the older offset methods. New machines for the process can fill a mold either directly or indirectly as in offset molding. Choice of operation depends solely on the type of mold used in the process. Offset molds are generally used when parts with inserts are called for. In this case, gates leading into cavities can be arranged so that molten plastic entering a cavity at high velocity will not disturb the position of inserts set up on delicate pins.

An operating difference which exists between the older offset process and the new transfer method is also worth noting. Cull removal in the offset process was accomplished by ejecting the cull while it was still attached to the part. In the new transfer-molding machines, the cull is separated from the part as the plunger moves back to start a new cycle. Thus, a part is ejected which seldom requires more than perhaps light flash removal for final finishing. Plunger or transfer injection methods are thus best suited to molding around inserts or to molding intricate shapes with combined thick and thin sections, particularly when close tolerances are demanded.

Screw-Type Machines

This discussion has thus far centered almost entirely on plunger-type machines. However, in injection molding of both thermosetting and thermoplastic resins, a major problem is that of producing a homogeneous melt which also entails good color dispersion and controlled viscosity. In the early days of injection molding, other industries were externally mixing and extruding ropelike, thoroughly homogenized, material to be charged into transfer pots (58–61). Subsequent progress in the design of injection-molding equipment led to the development of plasticators, spreaders, torpedos (Figs. 1 and 5), melt extractors (Fig. 9), and other devices to promote melt homogeneity. Unfortunately, many of these designs were not very successful with either highly viscous melts or materials which degraded at temperatures close to their melt temperatures, eg, rigid poly(vinyl chloride) (62).

By 1950, however, screw preplasticators were in use, some as "piggy-back pre-plasticizers" and others in horizontal and vertical arrangements where the injection cylinders were located at 90° to the plasticators. Long before the advent of "in-line" screw injection molding, modified screw preplasticators made possible the molding of rigid vinyl pipe fittings and strainfree, heavy sectioned acrylic moldings (63–67). The new machines (developed around 1966) do not require preforms or separate preheating; cold thermosetting compounds, for example in powdered form, can be fed directly into the hopper of the screw barrel. The extruder-type screw, in turn, feeds into a transfer chamber or pot.

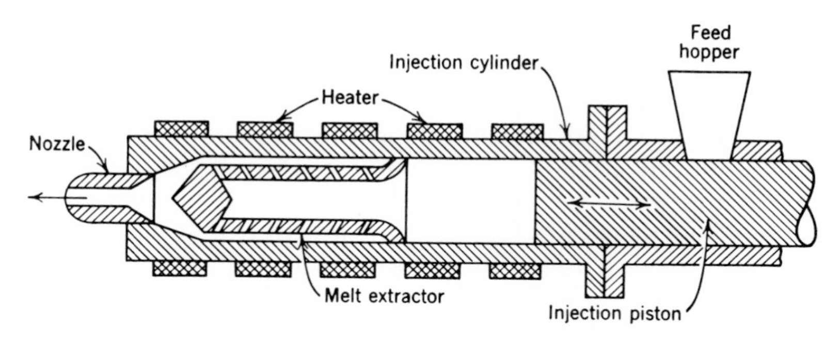

Fig. 9. Plunger–melt extractor plastication system. Plastic is forced through the core of the extractor and out of multiple radial holes into the annular space between the body of the extractor and the inner wall of the cylinder.

Screw transfer injection molding of thermosetting resins is able to operate at higher stock temperatures (Fig. 10) than the older plunger method because the powdered material can readily be brought to a melt which, in turn, can be rapidly moved into the transfer pot. Because of the higher preheating temperatures (250–280°F) very fast cure cycles are attainable. It is possible in some circumstances to mold thermosetting materials by this method at speeds as fast as thermoplastic resins can be cycled in ordinary injection molding.

In all, several types of transfer injection machines are in use, some are horizontal, single plunger units, equipped with automatic preform handling and dielectric heating means; these have largely replaced older offset machines (p. 55). Some transfer injection machines are plunger-plunger units and others, as previously described, are stationary screw-plunger combinations; yet others use a reciprocating screw preplasticator and transfer plunger to effect molding. All of these transfer injection-molding machines, except those axial units which use preforms, are bottom pot transfer arrangements where the plunger may act as a transfer plunger or become the force plug of a single-cavity mold (Fig. 11) and transfer plunger in one (see also Molds). Here, as has been noted, the preplasticator may be either a plunger or screw type.

Multiple-mold, rotary-type transfer or injection-molding machines and other highly specialized machines which use an injection ram (reciprocating screw or plunger) have been designed and built. Several of these rotary machines (Figs. 12–14) have specific applications in the molding of rubbers and elastomers.

Reciprocating-Screw Injection Molding. A plasticator as a separate unit may be used to feed a homogeneous melt into another unit such as a transfer pot used in transfer molding, an open mold used for casting or a closed mold used in compression molding. It should also be clear that a screw extruder, for example, can be

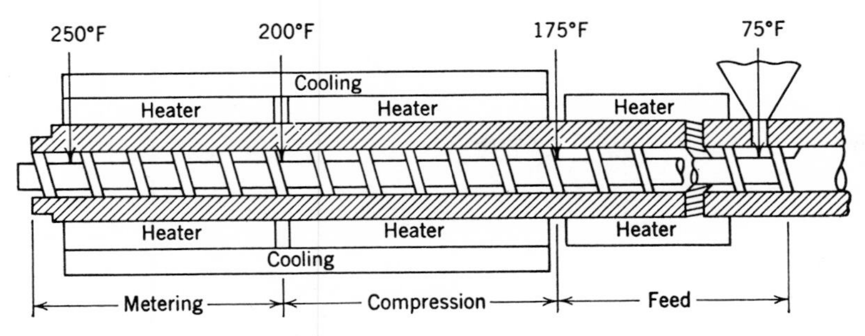

Fig. 10. Temperature gradient typical of plasticated material in the barrel of a screw transfer machine during high-speed molding of thermosetting resins.

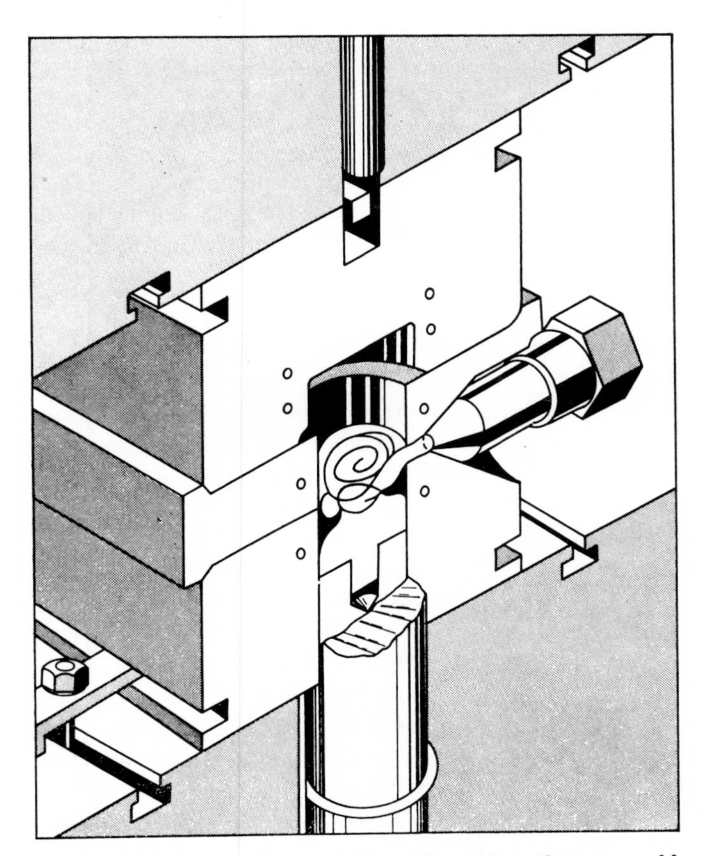

Fig. 11. Preplasticated material being forced into the open mold.

mounted to discharge directly into an injection cylinder, or an accumulator, or even into the parting line of a mold held in a clamping device. Although such devices have been in use for twenty-five to thirty years in highly specialized operations, they have mainly been special machines held captive by manufacturers not directly connected with the plastics industry and with injection molding. Molded footwear (68–70) and encapsulated electrical and electronic equipment have been processed on these complex injection-like machines.

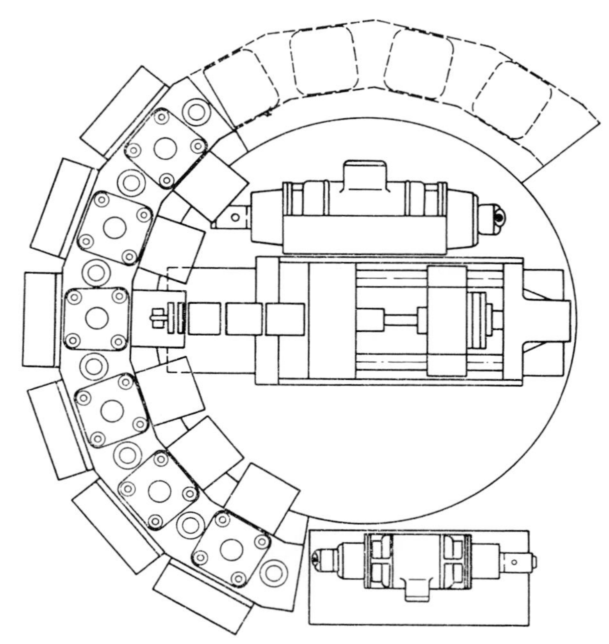

Fig. 12. Rotary injection-molding machine. Plan view of screw injection cylinder mounted on a turntable. Table indexes so that each mold set, clamped vertically, is fed by the injection cylinder along the parting line of the mold. Courtesy Eckert and Ziegler.

Although screw plasticators, as they were combined in these machines, may appear almost indistinguishable from the new reciprocating-screw injection units, they are in fact different in one important aspect: operating pressure. The screw extruder that was previously used is a relatively low-pressure injector, operating at pressures on the order of less than 5,000 psi, whereas the reciprocating-screw injector is a pressure device operating at pressures in the range of 10,000–20,000 psi. The incorporation of an internal sliding-ring or ball-type check valve (also termed nonreturn valves) affixed to the pump end of the extruder screw (71) enables the reciprocating-screw injector units to generate such high pressures (Fig. 15).

The Screw. In most cases the reciprocating screw used in injection molding is a simpler design than a melt-extruder screw (Fig. 16) (see also MELT EXTRUSION). The majority of extruder screws have an L/D ratio of 20:1 or more. Injection screws may be shorter, with ratios comparable to extruder screws for rubber. Screw length has been limited mainly because of the mechanical problems encountered in extending the machine. Moreover, it has not yet been established that the high ratios of extruder screws are needed for injection (72). But all other specifications, such as compression ratio, channel widths, and helix angles, required for optimizing the mixing and/or output of specific polymers are certainly applicable (73). It is interesting to note that compression ratios for thermoplastic resins range from 2:1 to 5:1 with the mean occurring at about 3:1. Thermosetting resins, on the other hand, are generally run with screws having compression ratios of approximately 1:1 or a maximum of 1.5:1. It is rare indeed when the particle size of thermosetting resins is big enough to require a compression ratio of 2:1. (Compression ratio is the ratio of the screw-channel

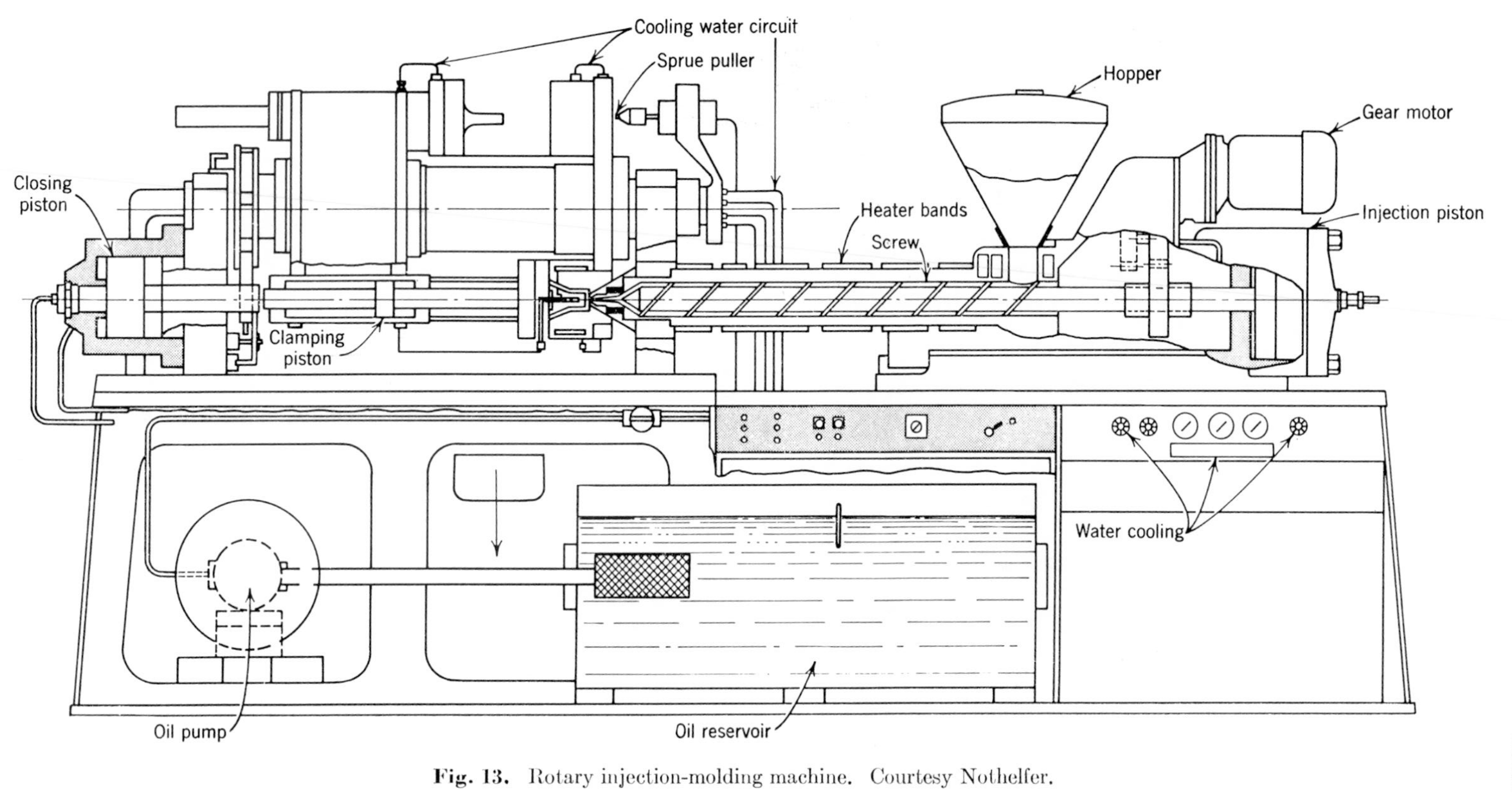

Fig. 13. Rotary injection-molding machine. Courtesy Nothelfer.

Fig. 14. Rotary injection-molding machine. Courtesy Husky.

volume at the feed end to that at the metering or pump end. It is designed not only to compact the material, but also to conform to the rate of melt and volume change as the material passes from a solid to a viscous fluid.)

Compound screws with decompression sections (used for degassing and devolatilizing) and those with dulmage mixing noses (ie, those with a smearing action) are not used for intermittently rotating injection screws. However, there are machines which employ continuously rotating screws; these, of course, may adopt all of the screw designs developed for melt extrusion. Those screws which are not equipped with check valves or wiping (smear) heads (Fig. 17), which completely fill and follow the contour of the cylinder, often develop dead spots of stagnant material and/or pigment. Twin-screw units, in particular, have a tendency toward producing dead spots.

Interaction between the rotating screw and cylinder brings the solid material to a melt and then pumps the melt forward. Friction between screw, material, and cylinder provides part of the heat necessary to bring the powders or pellets to a plastic melt. The amount of heat generated is contingent on the nature of the material, the type of screw, and the screw speed. Continuously rotating screws reach a steady-state condition when heat losses, by convection in the melt, and through the screw and cylinder by conduction and radiation, balance the heat input due to friction. A cylinder is cooled when temperatures are in excess of what is needed; it is heated initially to melt the material, and also to act as a blanket and thus reduce conduction and radiation losses to a minimum.

As the screw rotates, the heated (molten) and unheated (solid) layers of plastic are mixed and mulled in the screw flights as they are conveyed toward the nozzle end. Forward movement can only occur when friction at the screw channel is less than at the cylinder wall. When the converse is true, no pumping action is possible, the melt merely rotates with the screw and will not move forward.

Design Features and Injection Cycle. The main features of a reciprocating-screw injector (74–90) are shown diagrammatically in Figure 18. A screw rotates in an appropriate cylinder which has a hopper inlet and nozzle outlet. Material fed into the hopper is transported by screw rotation toward the nozzle where it fills a space or chamber. Pressure of the material advancing along the screw and filling the storage

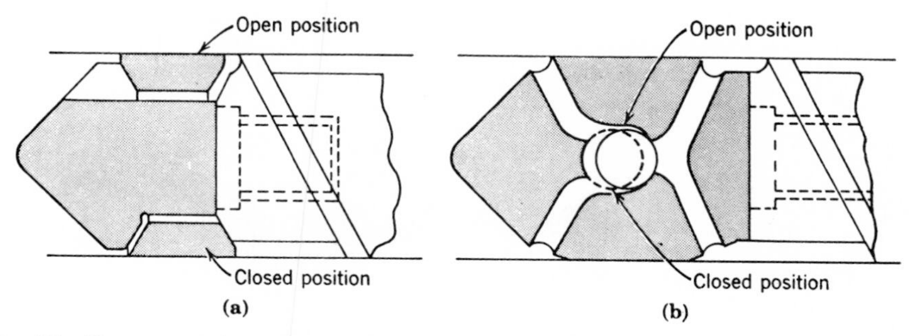

Fig. 15. Two types of nonreturn valves used on the front end of reciprocating screws in injection machines to prevent back flow along the screw channel during the injection stroke. (**a**) Check ring design; (**b**) ball check design.

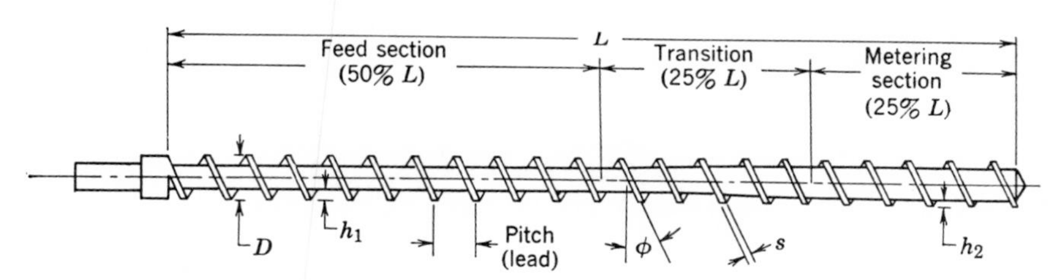

Fig. 16. Metering-type screw, showing recommended proportions for feed, transition, and metering sections to be used in screws for injection-molding machines. The sketch also defines the principal screw dimensions and critical parts. *D*, diameter; φ, helix angle; *s*, land width, h_1, flight depth (feed); h_2, flight depth (metering); *L*, overall length.

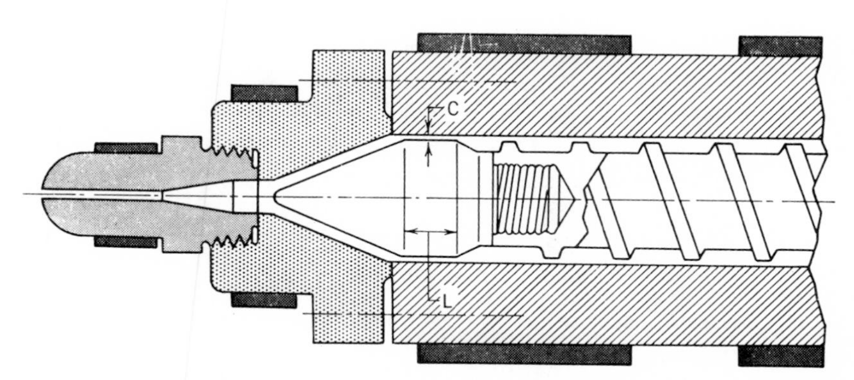

Fig. 17. Typical design of smear head for reciprocating-screw injection machine. Clearance (C) and land length (*L*) are made small enough and large enough, respectively, to prevent back flow while simultaneously promoting higher shear.

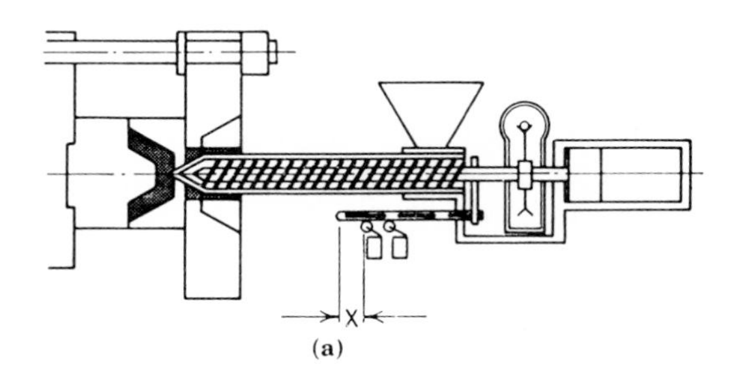

(a)

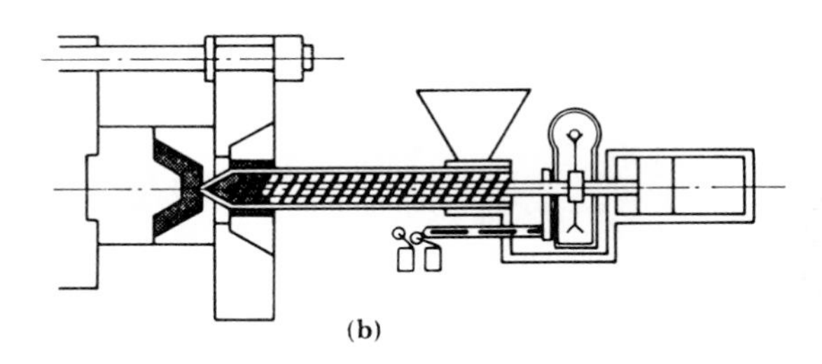

(b)

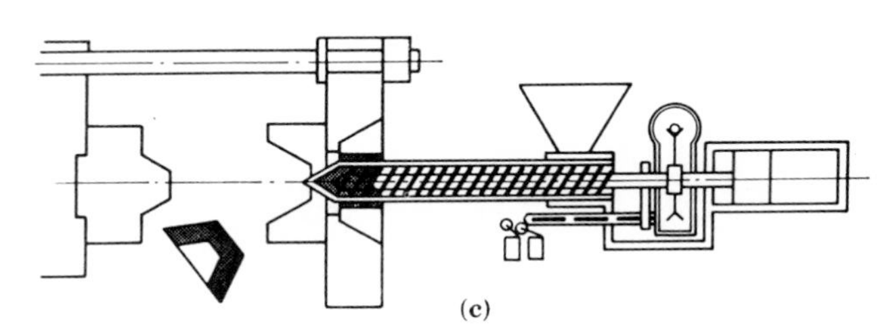

(c)

Fig. 18. Schematic diagram of a screw injection-molding machine. (**a**) Mold closed and locked; material already injected. (**b**) Mold is filled and remains closed; screw rotation causes material accumulating at front end to move screw backward. (**c**) Material for the next injection shot is ready. The molded part is ejected.

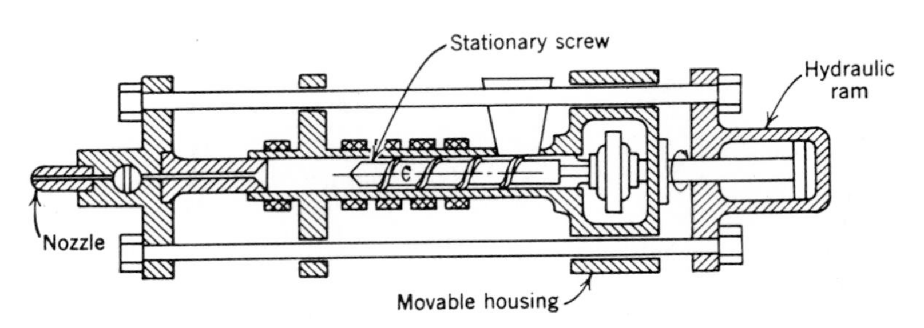

Fig. 19. Schematic diagram of screw injection machine with reciprocating housing.

chamber causes the screw to move backwards until the chamber is filled and/or the screw reaches the end of its travel to the rear.

In a typical screw injector, the actual injection is done by pushing on the rear of the screw, with either a separate or coupled hydraulic ram (91). The screw itself acts as the injection plunger, and is returned by the pressure of the plastic accumulating in front of it, so that the material is kept under continuous pressure.

Many variations of the basic reciprocating screw injection machine exist. In one machine, the screw is held stationary while the cylinder and its entire housing are caused to move back by the pressure of the plastic melt accumulating in front of the screw (Fig. 19). In another, the screw is also caused to retract but the machine makes use of several injection strokes to fill an accumulator cylinder. The heated cylinder with its accumulated charge is then used to fill larger-capacity molds than could be filled with a single stroke of the machine.

Plastication

In plastication a solid material is converted to a molten, viscous liquid. In injection molding, the process is carried out in the confines of the injection cylinder (Table 1) where plastics are heated and thus converted into a plastic melt of controlled viscosity (92). The molten material can then be forced into a mold by a plunger working inside the cylinder (93–98). Once the melt is trapped in the mold, it is re-solidified by either the application of heat, as in the case of thermosetting resins, or of cold, as required by thermoplastic materials.

Table 1. Relative Plasticating Capacity of Thermoplastic Materials in Screw and Plunger Machines

Plasticating capacity, lb/hr	Shot volume, cu in.
Plunger type	
20	4
40	10
60	20
70	40
Screw type	
28	4
60	10
90	20
130	40

The injection cylinder is basically a simple heat exchanger. It has heavy steel walls designed mainly to support the high internal pressures used in the process, and which, as a consequence, also act as heat sinks helping to hold temperatures constant (99–102). Internal surfaces are highly polished to reduce the amount of friction between plunger, material, and cylinder wall when the melt is forced through the cylinder into the mold.

Cylinders may be lined with corrosion-resistant sleeves which resist deterioration from degradation products of thermally unstable resins (see also DEGRADATION). Hardened liners of metal carbides and borides, sintered, ground, polished, and shrink-fit into steel cylinders, are used when highly abrasive materials or corrosive chemicals are present in the melt. Heat-treated or nitrided, and hard chrome-plated surfaces are also used for general-purpose molding compounds.

The most common means of heating a cylinder is by electrical resistance. Replaceable band-type heaters, fitted tightly about the periphery of the cylinder, are spaced along its length and divided into different heat zones. Demand for heat is not the same along the length of the cylinder. Generally, for thermoplastic resins a large heat input is required at the back where the cylinder contacts the water-cooled hopper feed sleeve; for the rest of the cylinder, however, heat input and control are contingent on the type and design of the plasticator. The opposite is required for thermosetting resins, ie, less heat at the hopper and more at the front of the cylinder.

The rheological properties of a material determine whether or not external heat is necessary to melt the material. Relatively shear-sensitive materials become considerably less viscous as they pass through the nozzle. For example, some nylons, although hard and rigid in the solid state, form low-viscosity fluids when melted. In this case mechanical shear is ineffective in generating heat, and conduction through the cylinder walls must be relied on to melt the polymer. On the other hand, for polymers such as polyethylene that can undergo considerable deformation, mechanical working in an extruder is an effective means for producing the heat necessary for plastication. In this instance, external heat is generally required only for starting an operation or when the material in the cylinder has been allowed to harden during a period when the machine was shut down.

In plunger-type machines, plastication commences when the charge of pellets or powder is pushed into the confines of the cylinder. Heat is transferred to the material through the walls of the cylinder, and is generally removed from the point of entry to the cylinder. As previously mentioned, the feed section is usually cooled to avoid premature softening or the possibility of material bridging or "hanging-up" as it flows from the hopper.

The low thermal conductivity characteristic of polymers in general and the thermal instability of some resins are both disadvantages to efficient heating; several methods have been devised to overcome these difficulties. Early designs for achieving a melt essentially free of hard gels and unmelted particles used a spreader or torpedo to reduce the thickness of the material layer and thus to facilitate heat transfer. Some machines also operated with heated torpedos to further increase the amount of heat which could be transferred to the material as it moves through the annular space between the torpedo and the cylinder wall. These devices (see Figs. 1, 5, and 9) are found in all plunger-type injection-molding machines for thermoplastics. (Machines used for molding thermosetting resins cannot use spreaders or any other mechanisms which offer resistance to flow, since premature hardening of the polymer in the cylinder must be avoided.) This resistance in moving granular thermoplastic material between the torpedo and the cylinder walls absorbs a great deal of power. Further impedance to flow is caused by the spider supports which center the torpedo in the cylinder bore. These restrictions plus nozzle orifices and mold gates cause the large pressure drops between the cylinder and the mold cavity.

Preplastication. Plunger units described thus far are suitable for molding items which require moderate amounts of material. However, as the cylinder bore grows larger in the larger-capacity units, the time of plastication increases such that few materials can be processed without being degraded. In other instances, the increased distance from the heat source to the material causes the pellets to melt unevenly. In such cases preplasticating units are used in which plastication is carried

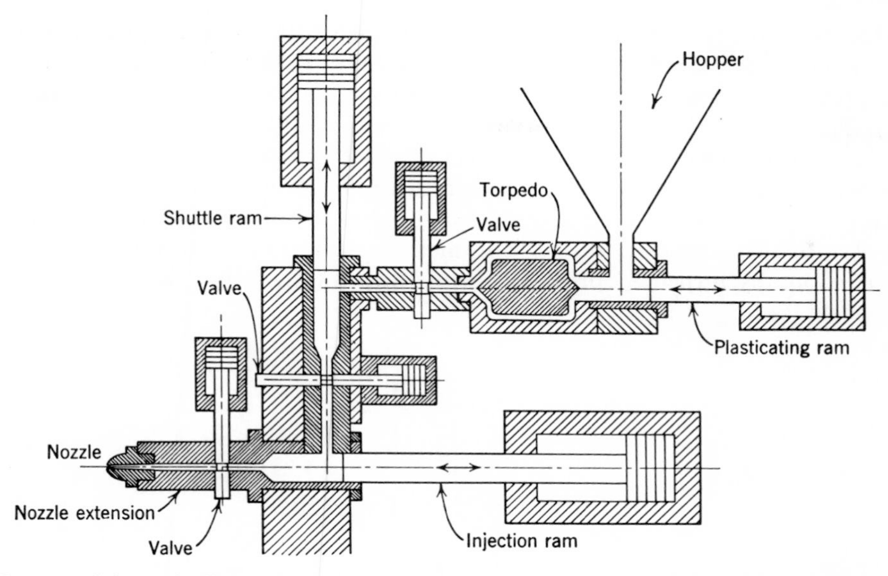

Fig. 20. Schematic diagram of plunger–plunger (cascade) preplasticating injection machine. Courtesy HPM.

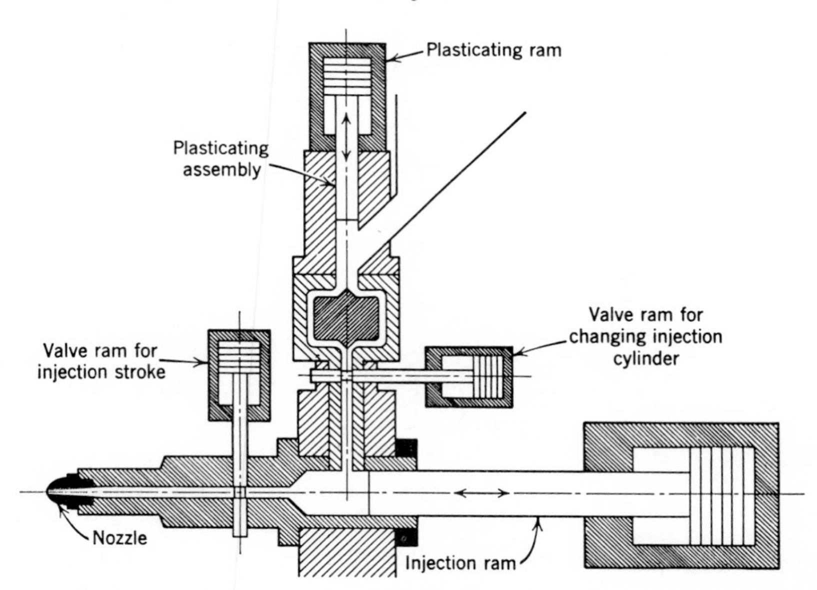

Fig. 21. Schematic diagram of plunger–plunger preplasticating injection machine. Courtesy Crown Machine and Tool Co.

out in a separate plasticating cylinder which feeds molten material into an injection cylinder. In some designs, the change from a solid to a liquid is effected by passing polymer particles through two or more plasticating cylinders (Fig. 20) (arranged in cascade formation) which are either plunger units equipped with spreaders or extruder screws (Fig. 21). The plunger–plunger machine functions in similar fashion to the older, plunger-type extruders.

Although plunger–plunger preplasticating machines have a capacity sufficient for molding large "shot" sizes, these machines, in common with single-cylinder machines, require other aids to effect color dispersion and a measure of melt homogeneity. Dispersion aids built into nozzles (see the section on Nozzles) were designed to introduce turbulence into the otherwise laminar flow of the melt. Recently machines have been fitted with screw-type preplasticators (103). Here, kneading of the material is sufficient to bring about color dispersion while simultaneously working the melt into a homogeneous mass. This method is superior to conventional methods of plastication by torpedos, spreaders, melt extractors, and other aids used in plunger-type machines.

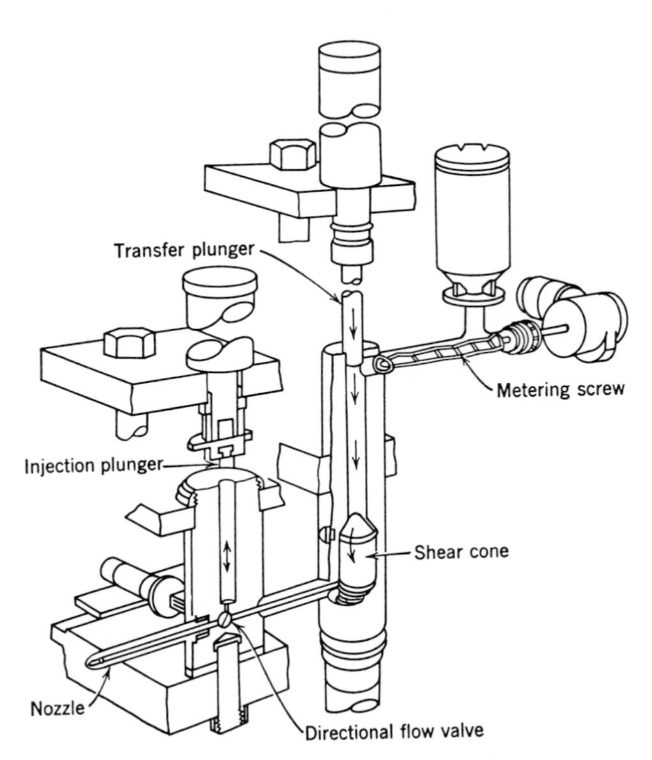

Fig. 22. Rotating shear-cone preplasticator.

Variations in granular feed have no effect on the injection process when preplastication is used. Lower pressures may be utilized in the injection cylinder to pack a mold with melt with greater precision and control. In addition, a longer, controlled heat input generally ensures a more uniform melt which, in the case of the screw preplasticator, is also one of greater homogeneity.

Disadvantages of preplasticator machines include the inconvenience of changing colors or materials. More material is normally required to purge these machines than is required to purge either the single-plunger type or the in-line screw extruder machines. The increased residence time of the material in the cylinder and the possibility of material sticking or hanging up in valving are also disadvantages.

A recent design in preplasticating machines makes use of a rotating shear cone plasticator rather than a rotating torpedo or screw (Fig. 22). The mechanism operates on the friction shear principle to plasticate and homogenize the material. Plastication takes place as a plunger moves the plastic material through an annulus formed between the rotating cone and housing. An inherent characteristic of this system is that melting of the resins takes place at minimum temperatures.

Nozzles

The nozzle of an injection-molding machine is the connection which allows molten material to transfer between the injection cylinder and the mold. It must be able to support the very high pressures encountered in injection molding while maintaining a good seal at its junction with the mold. Because the nozzle is a connector, several things are required of it. Its length must accommodate the space between the cylinder and mold. Its mass must be great enough to act as a heat sink. And, when under

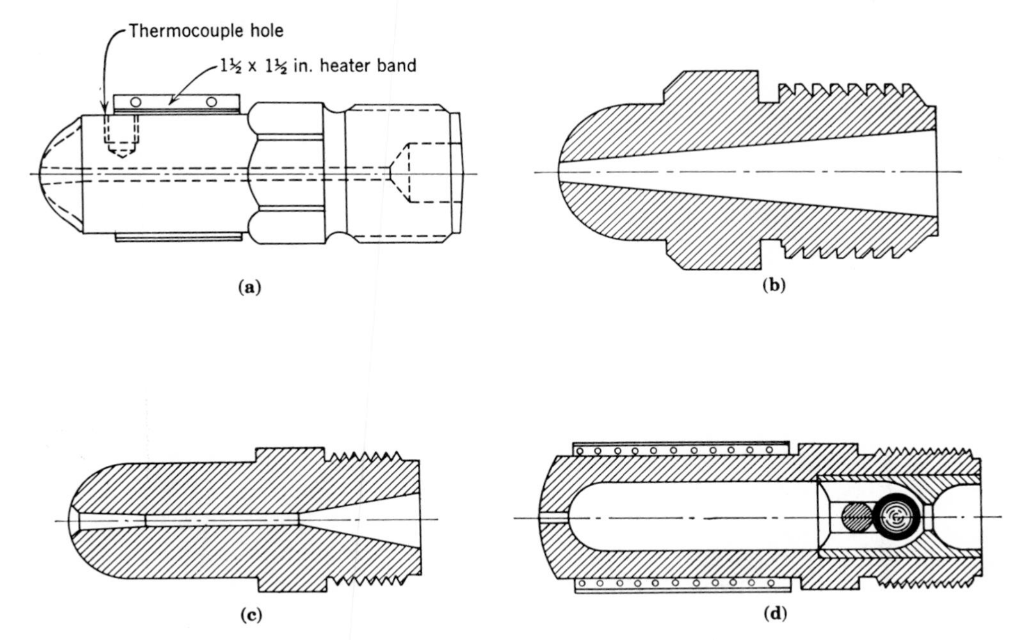

Fig. 23. Nozzles for injection molding. (**a**) Heater band and thermocouple well for temperature control. (**b**) Standard thermoplastic nozzle. (**c**) Reverse taper for thermosetting resins. (**d**) Free-flow nozzle with ball check valve.

static pressure, it must not allow material to drool or ooze out of the machine. In effect, the nozzle must accommodate the mold and, conversely, the mold must be made to suit the nozzle. Nozzles are commonly considered as tooling, in the same sense as molds. They are so closely allied to molds that some types must be designed with the mold in order to function properly.

There is a large variety of nozzles, ranging from the simplest, smoothly tapered nozzles to valved types which include automatic, hydraulically actuated shut-off and check valves (Figs. 23 and 24). In fact, they are as numerous and individualistic in design as are the molds which are used in injection molding. Some of the more important types will be described here; others may be found in the books listed in the General References.

Nozzles are generally made of tough, hard, tool steels; they can also be made of beryllium–copper. They are built with sufficient mass and heat capacity so that they cannot cool rapidly. Rapid cooling could cause material to harden in the orifice of the nozzle where, as a hardened plug, it can contaminate the next shot into the mold or cause other difficulties such as plugging and not allowing the mold cavity to fill. This has even been known to happen in molds with built-in cold-plug wells.

Nozzles for Thermoplastic Resins. Nozzles with bullet-shaped noses are the most prevalent types; these operate in conjunction with concave sprue bushings.

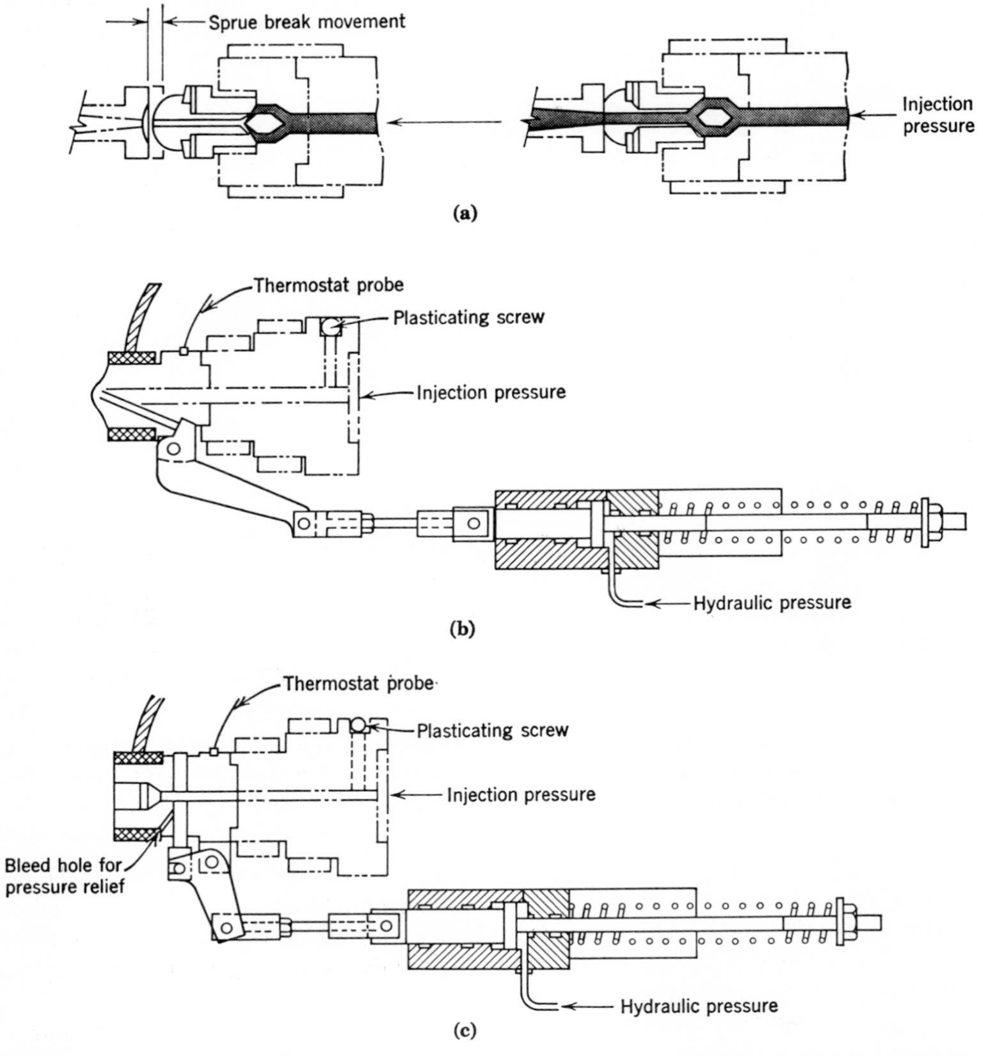

Fig. 24. Operation of nozzles: (**a**) precompression nozzle; (**b**) hydraulic nozzle with integral tip; (**c**) hydraulic nozzle with pressure relief and changeable tip.

The radius of the sprue bushing indentation is made larger than the radius of the nozzle so that contact is limited to a small arc of the sphere. Two reasons are given for this shape. First, the smaller the contact area between the hot nozzle and cold sprue, the less heat can be transferred from one to the other. Secondly, spherical contact surfaces are able to effect a seal even though axial misalignment occurs or dirt particles lodge between bushing and nozzle.

Although they are less desirable because of alignment problems, flat nozzles are also used, especially in designs where the cylinder is made to shift sideways, which displaces the nozzle orifice from the sprue opening to shut off the flow of material.

Nozzle openings leading to the sprue are made smaller, usually by about $\frac{1}{32}$ in., than either the opening in the sprue bushing or, when a bushing is not used, the hole in the mold. This is to ensure that no undercut is present which will cause the sprue to stick and tear or to be retained in the sprue bushing when it should pull loose during ejection, even when the opening in the sprue bushing and nozzle are not perfectly aligned with each other.

The straight-taper nozzle, which tapers from a small opening at the back, was for many years considered to be the general-purpose or standard nozzle. Because of the large pressure drop encountered in its use, it is no longer used as frequently (104). To overcome this high resistance to flow, a modified version in the form of a straight, large bore is used, with well-rounded corners at the base, where it terminates at an orifice whose length is generally limited to 2.5 diameters.

The reverse-taper nozzle (Fig. 24**c**) was originally designed to process nylons or other resins which undergo a sharp decrease in viscosity when heated to melt temperature. Such resins usually solidify readily when heat is removed. Materials with sharply defined melting points tend to ooze from very hot nozzles and, conversely, tend to solidify in nozzles which lose heat to cold molds. As with thermosetting resins and elastomers, for which reverse tapers are also required, nylons and polypropylenes require a similar orifice because of their tendency to harden in the end of the nozzle adjacent to the cold mold. The reverse taper in this case facilitates removal of the solidified material attached to the sprue when the sprue is ejected from the mold and allows the nozzle to be run at a low enough temperature to avoid oozing.

Color Dispersion. Various aids to color dispersion may be incorporated in the injection nozzle (105), especially during the molding of polymers with sharply defined melting points such as polypropylene and nylon. In such cases, in plunger-type machines, resin nearest the walls of the injection cylinder melts first and flows with little or no turbulence. The molten polymer is subjected to greater shear, which in turn raises its temperature, so that it becomes even less viscous; the material away from the walls, being less fluid, has no tendency to mix with material near the walls. This flow pattern persists to a considerable degree even when various dispersion devices such as torpedos, melt extractors, and others with complicated and tortuous passages are used.

Several types of color dispersion aids exist which are fitted into nozzles. These aids to color dispersion consist of perforated breaker plates and inserts in the form of screws, and spherical venturis which operate in a way that is opposite to that in which a venturi throat operates. Where a true venturi smooths out and streamlines a flow, the spherical venturi introduces violent turbulence into a stream. Turbulence occurs at the entrance to a spherical pocket and again at a point where resin leaves the pocket (Fig. 25).

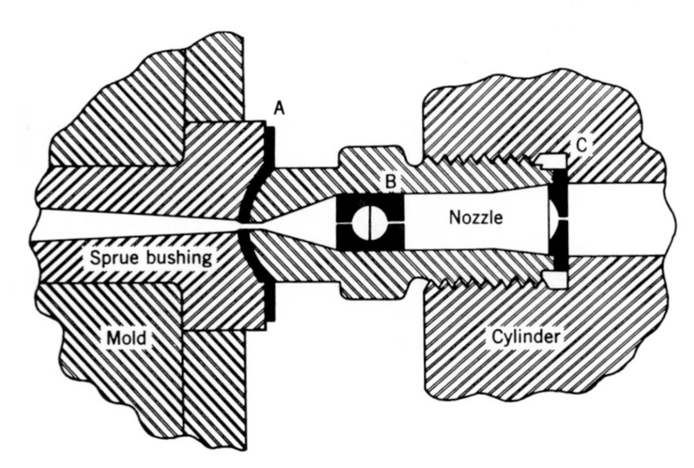

Fig. 25. Color dispersion aids. *A*, Orifice plate between nozzle and bushing; cannot be used when nozzle withdraws from bushing. *B*, Spherical venturi (see description in text). *C*, Orifice plate clamped in place by the nozzle.

The greatest drawback to the use of dispersion-nozzle inserts is that they always require higher injection pressures. Many, such as the spherical venturi with 0.020-in. diameter orifices can, for example with vinyl resins, require as much as 50% additional pressure to overcome the large pressure drop across the small orifices and yet maintain adequate pressure to fill the mold. Color dispersion aids are not normally used in nozzles for screw injection molding. Here, the rotation of the screw is generally sufficient to effect an intimate mixture of polymer and pigment for good color dispersion.

Automatic Shut-Off. Shut-off nozzles are used to prevent nozzles from leaking into empty molds or when the nozzles are retracted from sprue bushings. Some resins, such as nylon and polystyrene, become extremely fluid at their melt temperatures, to the extent that they tend to drool out of openings. Pressure in an injection cylinder is normally nil when leakage occurs; only the head of molten material causes leakage flow.

Valves of all kinds, ie, spring-loaded ball check valves, hydraulically and mechanically operated needle valves, and rotary valves, have been incorporated into nozzles. Among the most successful shut-off nozzles are those based on needle-valve designs. With this type of valve, the small plug which forms forward of the closed valve becomes part of the sprue and is removed with the sprue when it is ejected from the machine.

Valve gating nozzles used in sprueless molding also belong in the general category of shut-off nozzles, as do the nozzles which are used for impact molding.

Gating Nozzles. In the development of tooling in injection molding, mold gating nozzles (106), hot-runner (107,108) and insulated-runner molds, and nozzle manifolds preceded valve gating as means for obtaining sprueless moldings. Elimination of sprue and runner system reduces labor and the amount of scrap accumulated with each molding.

A gating nozzle, as its name implies, is a nozzle with a small opening which forms a gate at the junction of the nozzle with the mold cavity. When such a nozzle is equipped with a valve, it is known as a valve gating nozzle (109); without the valve, it is called a mold gating nozzle. In both cases, the nozzle takes the place of the sprue and runner system and shoots the melt directly into the cavity. Orifice diameters range from 0.020 to 0.060 in., which allow the molded part to tear free of the material remaining in a mold gating nozzle. Valve gating nozzles, on the other hand, close off and separate the molding from material in the nozzle.

Gating nozzles are also used to terminate nozzle manifolds. These units, usually in the shape of tubular structures, are attached through a "tee" section to the heating cylinder in place of the standard nozzle. Since a manifold is essentially an extension of the injection cylinder, it must be heated and occasionally also insulated, and must have its heat controlled by separate instrumentation. Uniform heat must be maintained on all legs of a manifold to prevent unbalanced flow.

Six-nozzle manifolds have been used with high-production sprueless molds in completely automatic molding operations. However, manifold arms, with more than six nozzles have not been used successfully. As small a number as eight nozzles on each arm of a six manifold section has produced "short shots" and moldings with excessive flash at peripheral cavities. The problem of temperature and pressure balance increases as the number of cavities in the mold increase and as the mass of each molding increases.

Impact Molding Nozzles. Valve gate nozzles operated by solenoid, hydromechanical lifters, or hydraulic ram lifters are used in impact molding. Almost all machines can be adjusted or modified for impact molding. For this molding method, a valved nozzle is programmed to remain closed while pressure is built up in the heating cylinder. Under high pressure, eg, 15,000–20,000 psi, a polymer melt can be slightly compressed. After the melt has reached a given pressure, the nozzle valve is opened; the molten polymer charge is forced through the nozzle orifice, rapidly filling the mold cavity.

Advantages gained by filling a mold rapidly are increased production and stronger and tougher molded parts. The latter results from the elimination of "frozen-in" stresses through the rapid and turbulent filling of the mold, which prevents orientation of polymer chains along flow lines.

Molds

An injection mold is a split mold clamped under pressure, with an orifice through which a heat-softened plastic material is injected into a cavity or cavities wherein the plastic material is allowed to harden before being ejected (110–121). A typical injection mold set is shown in Figure 26. Molds for injection molding are described in the article on MOLDS. Some special aspects of the subject will be discussed in the following sections.

Injection molds differ one from another more as the result of the product being molded than of the material being used, eg, thermoplastic resin, thermosetting resin, or elastomer. In general, orifice sizes are larger for more viscous materials and smaller for less viscous materials; they are larger for heavy-sectioned parts and smaller for thin-sectioned parts. Refinements which may be added to the basic mold components include ejector mechanisms, cores, unscrewing mechanisms (see MOLDS), as well as separate runner plates in three-plate molds (Fig. 27) and sprueless molds. One type of the latter is a modification of the separate runner plate, and is called an insulated-runner mold (Fig. 28). Others are the hot-runner mold (Figs. 29 and 30) and the multiple-nozzle manifold. Of the two designs, the multiple-nozzle manifold is easier to purge than the hot-runner arrangement.

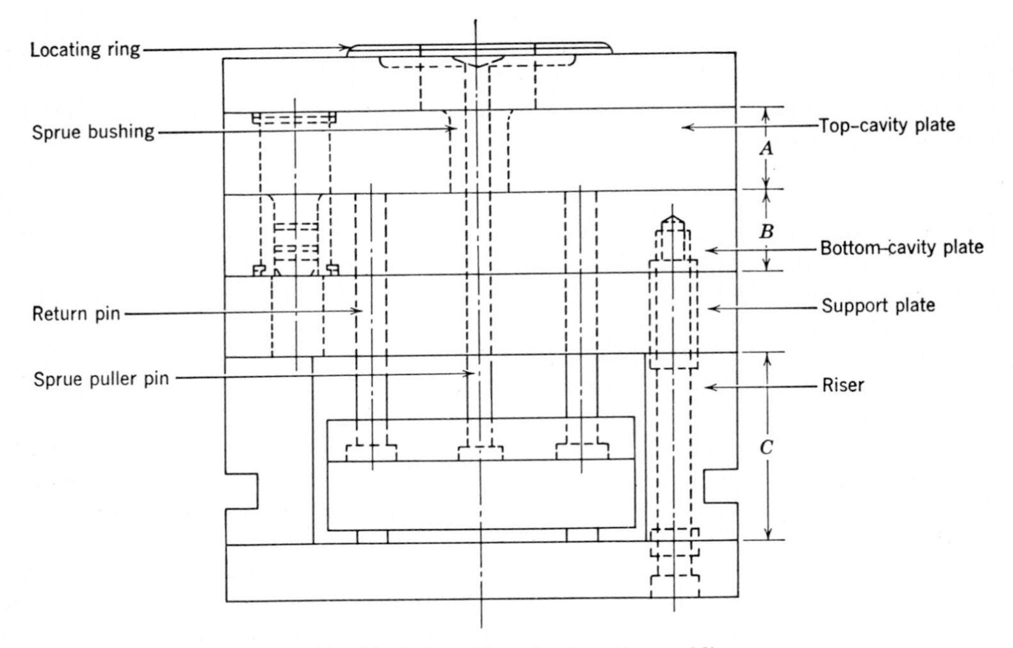

Fig. 26. Typical mold set for injection molding.

Sprues. A tapered sprue is generally used in injection molding. Tapered sprues may be machined directly into the stationary half of a mold or they may be inserted in a mold in the form of a ground and polished bushing (122). Sprue orifice sizes are always larger than nozzle openings and range from about ⅛ to ⅜ in. in diameter. They taper outward toward the cavity or runner system at a minimum included angle of 2°. The amount of taper is optional; however, the larger the taper, the easier it is to remove a sprue from the mold or bushing. Included angles of 8° and greater have been used; the angle is generally predicated on the characteristics of the molding material and the size, cross section, and volume of runners or cavities.

A well at the large end of the sprue, located in the movable half of a mold, acts to receive the cool slug of material which emerges first from the nozzle at the start of the injection stroke. In this manner the cold slug is retained and kept from plugging small gates at mold-cavity entrances. The cold-slug well also serves to pull the sprue away from the nozzle when the mold is opened. Several designs exist for a simple sprue puller. Undercut grooves and fish tails are more readily adaptable to rubbery or soft materials such as polyethylene, plasticized poly(vinyl chloride), and elastomers. Rigid materials are more amenable to the Z-cut, or hook, cut into an ejector pin heat forms the bottom of the cold-slug well. When the ejector system operates, the modified pin is pushed completely out of the well so that the sprue is allowed to fall free of it. The well should always be larger than the large diameter of the sprue; otherwise it may fail to trap the slug.

Runner Systems. Three types of runners are currently in use: the full-round cross section, the modified half-round, and the trapezoidal section. Registration of mold halves is particularly important when full-round runners are used since any offset affects the perimeter of the cross section and, consequently, the flow of the molten plastic. Flow resistance is partly determined by the ratio of the periphery of the runner cross section to the area. The smaller the ratio, the less resistance to flow offered by the runner. The cross-section periphery to area ratio is smallest in a full-round runner. A half-round or trapezoid of equal area has a perimeter approximately 15% greater than a full-round section.

As the melt of uniform temperature and viscosity enters the sprue opening and the stream splits into the runners, any difference in length, cross section, or rate of heat transfer of the runners will affect the quantity of material reaching the various cavities (Fig. 31). Thermal balance in a mold is one of the most difficult things to predict during mold design and to achieve in construction. This is mainly due to the mechanical interference of ejector, guide, and core pins with heating or cooling channels of the mold. Asymmetrically shaped cavities also contribute to the difficulties encountered in designing for thermal balance; even when these multiple cavities are exact duplicates of each other.

Insofar as possible, runner paths should be equalized, both dimensionally and thermally, from the sprue to each cavity. Since it is very difficult to construct a runner layout which will function perfectly on its first trial, it is always wise to allow for minor

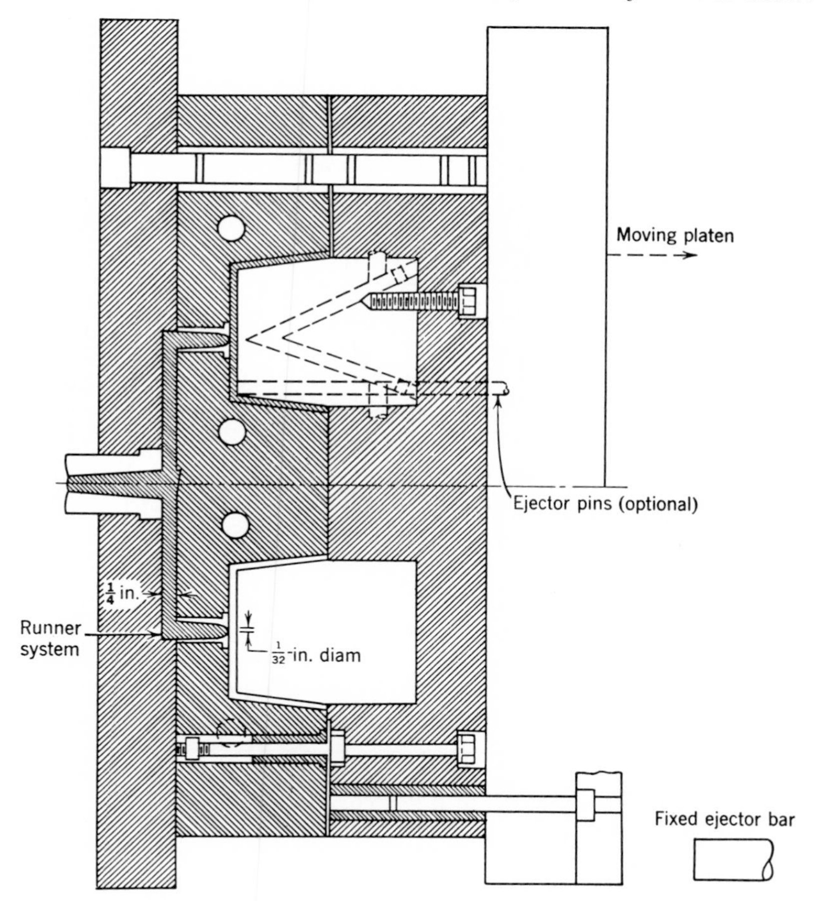

Fig. 27. Three-plate injection mold with pin gating.

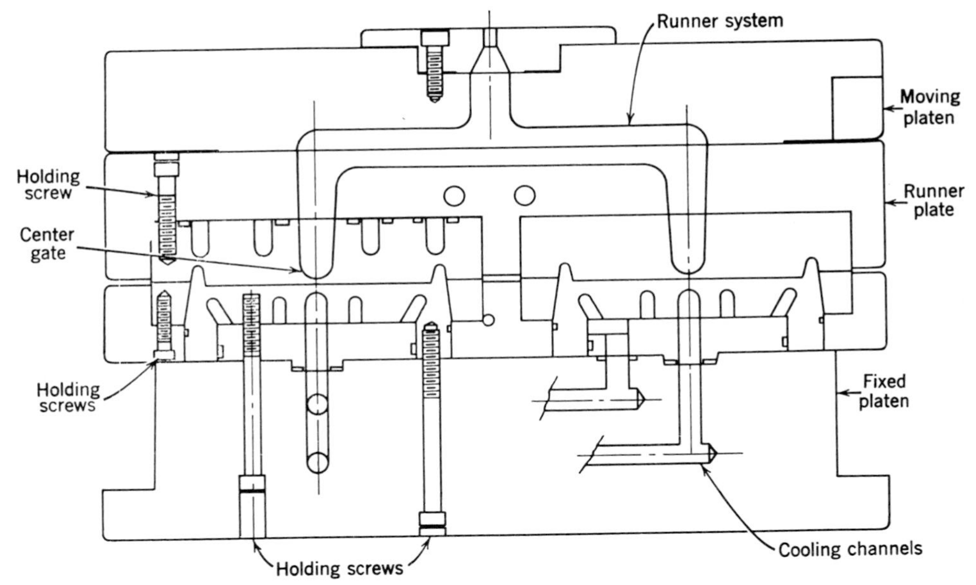

Fig. 28. Typical insulated-runner mold.

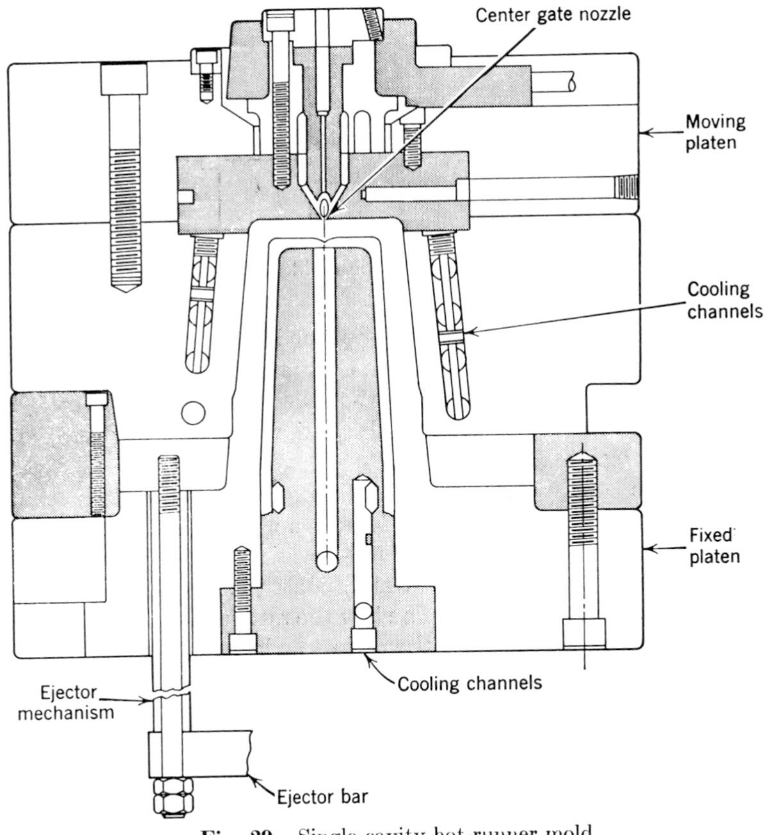

Fig. 29. Single-cavity hot-runner mold.

alterations to be made to the mold which will balance it after it has been test run. Alterations generally consist of increasing the width or depth of a runner channel.

Gating. A gate is the opening or orifice between the nozzle, sprue, or runner and the cavity in a mold (123). The various types of gates used in injection molding are described in the article MOLDS.

In a single-cavity, center-gated mold, the gate may be as large as the open end of a sprue bushing or smaller than the orifice in a nozzle. Center gating may be applied to both single- and multiple-cavity molds. A center gate that is also a pinpoint gate is

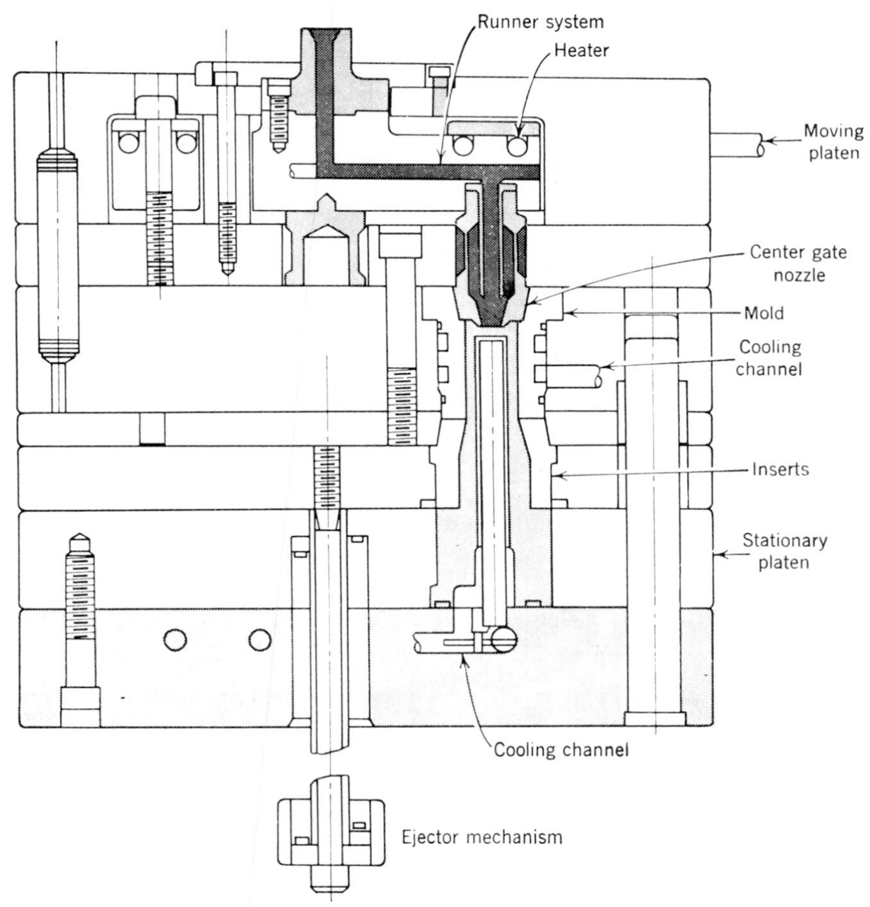

Fig. 30. Multiple-cavity hot-runner mold.

utilized in valve gating and mold gating nozzles, as well as in insulated- and hot-runner molds. Such gates have orifice sizes ranging from 0.015 to 0.060 in. in diameter. Modified pinpoint gates are also used for edge and submarine gating. Pinpoint gates cannot be used with highly viscous melts or materials which rapidly cure or freeze, or those that are susceptible to frictional degradation. They are generally relegated to the molding of small parts because the small volume of material instantaneously flowing through the gate tends to set up rapidly (which in a large cavity could be responsible for molding incomplete, or "short-shot," parts).

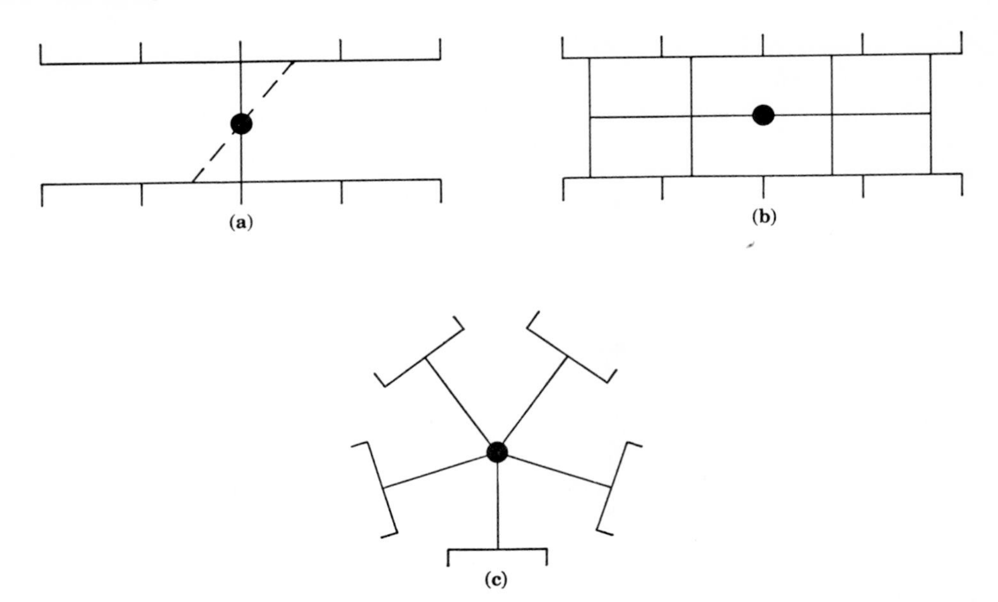

Fig. 31. Equalized runner layouts are easily devised when the number of cavities is some integral power of two, but may present problems when the number of cavities is a multiple of three or five. Layout **a** is the traditional way, with gross inequalities in flow paths, dashed line is slight alternate modification that would prevent direct flow into gates of central cavities, but still imposes inequity of flow, filling, and freezing times in the ten cavities. Layout **b** would give much better equalization but requires more total runner length. Layout **c** looks unconventional but provides the exact equalization needed when dimensional tolerances are tight, without increase in total runner length.

Gate sizes are based on wall thickness or size of the part being molded. Orifice sizes should be as large as practicable with minimum land length to allow rapid filling of a cavity and to reduce the effect of material surging into the cavity.

Where larger gates are required, fan or tab type gates are used, especially for molding those materials with high melt viscosities. Fan gates are generally used in edge-gating applications. Tab gates, on the other hand, are an extension of the runner over the edge of a part, resulting in a molded tab on the part. Such a tab may either be hidden in the design of the part or trimmed off after molding.

One of the most important aspects of mold design is the placement and the number of gates used in feeding a mold cavity. For example, a fan gate feeding a cylindrical part sends a stream of melt flowing in both directions about the center-hole core; a weld of the two streams occurs at the edge farthest from the gate. If the melt flows slowly, the welding of the two streams may be incomplete. At the very least, a weld line will appear on the molded part which may or may not be structurally sound. Such a part would require at least two gates or, if feasible, either a ring or disk gate.

Sprueless Molds. In injection molding of thermoplastic materials, hot- and insulated-runner systems are used whenever the regrind ratio is calculated to exceed 15–20%. (Regrind ratio is the weight of the runner and sprue system to the weight of the molded parts.) In molding thermosetting materials, runners and sprues are lost material (since they cannot be reused) and as such they must be considered in the overall cost of molding. Center gating and valved center gating on single-cavity molds have been successfully utilized in eliminating runners and sprues in thermosetting injection molding.

Hot- and insulated-runner manifolds are three-plate injection molds that are designed to allow for center gating of mold cavities. A hot-runner system, as its name implies, is a three-plate injection mold in which the runner plate is heated, usually with internal cartridge type, high watt density, electrical heaters. The hot-runner plate is isolated from the cold mold by an air gap or insulating materials such as rock wool or glass-fiber insulation. The same type of runner arrangement without the cartridge heaters is known as an insulated-runner system. Another type of insulated-runner system uses an enlarged runner cross section situated between the runner plate and the mold mounting plate. This arrangement makes use of the insulating property of the plastic which solidifies on the walls of the runners and narrows the channel through which the molten plastic flows. Both systems depend solely on the heat of the melt to keep the plastic material in a fluid state.

A third type of sprueless mold is a two-plate mold equipped with a nozzle manifold, as described in the section on Gating Nozzles (p. 71).

Sprueless molds are generally used in highly automated injection-molding machines equipped with mold or valve gated nozzles. They are used with both single- and multiple-cavity molds.

Ejection Systems. Injection molds are preferably designed so that molded parts are retained by and move with the movable force block when a mold is opened. In a simple two-plate mold this action ensures the removal of the sprue and runner system from the stationary mold half. Sprueless molds also use this action to tear the part loose from the mold or valve gate.

Three factors determine the proper ejection of molded products. These are polished cavity surfaces, the use of maximum taper that can be tolerated by the molded part (in the direction of draw) without affecting dimension changes of fitted parts, and the proper venting (124) of mold cavities. The latter facilitates the removal of the molded pieces from the force block without drawing a vacuum under them. Suction between a molded part and a highly polished cavity, as the part is being ejected, is a major factor in producing torn and distorted parts.

In most cases, roughening the surface of a mold in combination with 0.0002-in. grooves strategically cut at the parting line is sufficient to vent a mold. Other vent points normally occur in the clearance between ejector pins and their bored holes in the mold. Lenses and other highly polished surfaces which cannot tolerate the marks of ejector pins are a special case. Here, the entire surface of the lens becomes the ejector; the polished mold bottom is attached to the ejector plate, which moves it out of the cavity to eject the finished molding.

Shrinkage of plastics (125–127), which is considerably greater than that of metals, can also be a source of trouble in ejection. For example, a part can shrink so tightly around several core pins as to preclude removal without damage. Stripper plates around cores, tapered cores, and judicious use of mold lubricant (128) tend to relieve the sticking due to shrinkage.

Insert Molding. Inserts may be either molded in place or pressed into the hardened part after molding. When inserts are molded in they must first be properly anchored in the mold so that the inrush of molten polymer will not alter their location. Metals are the most common materials for inserts but glass, ceramics, and other plastics are also used. Unless the polymer is fairly flexible, the coefficients of thermal expansion of the molding compound and the insert material must be similar, nor can chemical reactions occur between the insert and the plastic, as, for example, between brass and poly(vinyl chloride). Furthermore, the plastic shrinks around an insert as it hardens, creating internal stresses which, on aging, will eventually be relieved by cracking of the part.

Loading fixtures used in horizontal injection-molding machines are generally of the same type as those utilized in compression and transfer molding. Vertical-platen machines, on the other hand, have used hopper-fed insert loaders in fully automatic operations. Most injection moldings which require inserts are molded on machines which are operated semiautomatically. Magnets are sometimes used to hold iron or steel inserts in a vertical mold.

Multicolor Molding

Injection moldings of two or more colors are utilized for typewriter keys, identification tabs, modern tableware pieces, and instrument housings. These are made by first producing a molding in one color and/or material, which becomes an insert in a second mold where another color or material is introduced to embed a portion of the inserted molding. Separate injection-molding machines may be used for these multiple operations. Special automatic machines have been introduced that are equipped with multiple cylinders and nozzles (129). A shuttle table on which molds are mounted completes this machine. Material from one nozzle is injected into the cavity and allowed to harden or cure before the table is indexed to receive the next injection charge.

For the machine to operate properly, the two cavities must be different in form. The fixed halves of both cavities must be solidly attached to the machine and remain stationary, and the index table with the two identical force plugs must then retain the molded part on the force or core which received the first injection shot. The finished part on the second core which received both colors can then be ejected. It is apparent that for such an operation to be successful, indexing must be sequenced into the automatic machine cycle and the initial molding, at least, must be sprueless. The arrangement must therefore be equipped with either mold or valve gate nozzles.

Auxiliary Systems

Mold Clamping. A clamp mechanism must keep the mold halves closed under the pressure created by the melt in the cavities, runners, and sprue. Clamping may be hydraulic, mechanical, hydromechanical, or electromechanical. Other less frequently used means include an arrangement of two opposed cylinders, each acting on one half of the mold, and a mechanical toggle which supports a hydraulic cylinder for the final clamping operation after the toggle links have closed and locked.

Temperature Control. *Thermoplastic* materials cool in the mold cavity by losing heat to the surrounding metal mold. Some polymers require a slower rate of cooling than others to avoid distortion; a slow cooling is also necessary for thick sections where surface skin will harden and trap molten material at the center. If the skin is not thick enough at the time of ejection, the part will shrink excessively and distort or harden with internal voids. When a slower rate of heat removal is called for, coolant flow through cooling channels may be reduced or the temperature of refrigerated coolant can be increased (130–135).

Injection molds for *thermosetting* compounds are generally heated by contact with electrically heated platens on which they are mounted. Besides the electrically heated platens, there are also electrically and steam-heated molds. These are heated and thermostatically controlled in much the same way that compression and transfer molds are heated. Cartridge-type electrical heaters placed in the mold are also used. These are generally placed in a balanced configuration so that heat zones are equal, and heat reaching the inner cavity surfaces is held constant. The body of the mold and the platen combine to form a large heat sink which tends to damp temperature fluctuations that occur from shot to shot. In other cases, live steam or heated liquid circulated through cooling channels in some types of thermoplastic molds may be used to adapt these to the molding of thermosetting resins.

Heaters. Heating apparatus for an injection cylinder for both thermoplastic and thermosetting molding should be flexible and adjustable to supply heat in a smooth temperature gradient from the hopper to and including the nozzle of the machine. Electric strip and cartridge-type heating elements are used predominantly in thermoplastic injection machines (136,137). Molding machines for thermosetting polymers may be equipped with either electric resistance heaters tempered by a circulating liquid coolant jacket, or simply with a jacket that circulates a heated fluid. In thermosetting molding machines, platen and mold heating can be by electric cartridge-type heaters or heated fluid (oil, water, or steam).

The most common type of heating element is the electric resistance unit. Other types include induction heating (138), coils, and dielectric antenna plates. The latter methods of heating are infrequently used in injection-molding machines and have not been in commercial use in the United States.

Instrumentation. Temperature is normally controlled by thermocouple-type pyrometer probes connected to recorder/controllers. Temperature-control systems used on injection-molding machines, include on–off, time proportioning, and saturable core reactor controllers. The most widely used temperature controller is the time-proportioning or anticipating heat control. The on–off instrument is rarely used since its time lag can cause excessive temperatures which might scorch sensitive materials. Saturable core reactors are less frequently used because they do not readily respond to changes in molding capacity which occur when mold sizes and molding materials are changed. In addition, they do not shut down completely when cooling is required. This type of controller is generally used on high-production machines built to process one type of material in a specific mold for almost the life span of the machine.

Bibliography

1. B. A. Olmsted, "Utility Requirements for Injection Machines," *Mod. Plastics* **43** (3), 131 (1966).
2. E. Simon, "Economics and Production Speeds of Medium Sized Injection Moulding Machines," *Plastics Inst.* **33** (12), 257 (1965).
3. R. W. Muhs, "Determining True Machine-Time Costs," *Plastics Technol.* **11** (8), 48 (1965).
4. O. Hartung, "Einfache und billige Spritzgusswerkzeuge für Kleinsterien," *Plaste Kautschuk* **18** (7), 414 (1965).
5. M. A. Sander, "Precision Injection Molding Through Statistical Quality Control," *SPE J.* **21** (7), 663 (1965).
6. T. J. Daniel and W. C. Hodson, "Productivity in Plastics Injection Molding," *Plastics Inst. Trans. J.* **33** (6), 83 (1965).

7. R. Sonntag, "Determining the Output of Injection Molding Machines," *Intern. Plastics Eng.* **5** (5), 195 (1965).
8. L. Jonup, "The Cost of Setting Up a Processing Line for Injection Molding," *Mod. Plastics* **42** (3), 95 (1965).
9. "Small-Run Injection Molding Can Be Economical," *Mod. Plastics* **42** (11), 90 (1964).
10. M. G. Munns, "Comparison of Pressure Diecasting and Injection Molding Equipment," *Plastics Inst. Trans.* **34** (112), 195 (Aug. 1966).
11. "Two Developments in Plastics Machinery," *Engineer* **222** (5763), 54 (July 1966).
12. "New Machines and Equipment," *Brit. Plastics* **39** (7), 404 (July 1966).
13. H. Landler and G. Stuis, "Trends in Injection Moulding Machine Construction," *Intern. Plastics Eng.* **6** (1), 1 (1966).
14. H. Castrow, "Einfluss der Wandungsneigung auf die Konstruktion eines Spritzgusswerkzeuges," *Kunststoffe* **55** (9), 731 (1965).
15. A. Smorawinski and H. Zawistowski, *Polimery* **10** (4), 166 (1965).
16. E. W. Vaill, "Technical Advances in Automatic Molding," *SPE J.* **23** (3), 60 (March 1967).
17. R. J. Lindsey, R. M. Norman, and W. B. Evans, "How to Get the Most Out of Screw Injection Machines," *Plastics Technol.* **12** (4), 31 (1966).
18. B. Meyer, "Materialzuschläge beim Spritzgiessen," *Plaste Kautschuk* **13** (3), 166 (1966).
19. D. Caspar, "Vergleichende Untersuchugen zu Druckmessungen in Spritzgiesswerkzeugen," *Plaste Kautschuk* **13** (2), 101 (1966).
20. J. Breitenback and G. Ohliger, "Neue Konzeption im Bau von Spritzgussmaschinen," *Plastverarbeiter* **17** (1), 9 (1966).
21. G. Holzmueller, "Probleme bei der Herstellung duennwandiger Plastteile im Spritzgiessverfahren," *Plaste Kautschuk* **12** (7), 416 (1965).
22. R. I. Mamedov and I. F. Kanavets, "Follow-up Conditions in Injection Moulding Thick-Walled Articles of Thermoplastics," *Soviet Plastics* **4**, 33 (1966).
23. A. G. Serle and P. R. Lantos, "Large-Part Molding of Acetal Copolymer Resins," *SPE Tech. Papers* **12** (3), 8 (1966).
24. J. K. Hark Leroad, "Molding Very Thin Sections from Polypropylene and Polyallomer," *SPE Tech. Papers* **12** (3), 3 (1966).
25. H. D. Bassett, "Injection Molding of Polysulfone," *SPE Tech. Papers* **12** (3), 3 (1966).
26. Z. M. Haycock, "How to Injection Mold PPO," *Plastics Technol.* **12** (2), 35 (1966).
27. E. A. Militskova, E. S. Viktorov, A. D. Sokolov, and V. P. Kostikov, "Injection Moulding of Polyformaldehyde," *Soviet Plastics* **1** (1), 28 (1966).
28. W. R. Schlich and R. S. Hagan, "Material Parameters for Injection Molding," *SPE Tech. Papers* **12** (1), 22 (1966).
29. V. Karpov and M. Kaufman, "Injection Molding of Glass-Reinforced Nylon 6/6," *Brit. Plastics* **38** (8), 498 (1965).
30. A. G. Serle, "Dimensional Behavior of Acetal Copolymer Molding," *Mod. Plastics* **42** (6), 115 (1965).
31. E. C. Jameson, "Injection Molding Technique for Large Parts," *Mod. Plastics* **42** (5), 165 (1965).
32. S. B. Neff, "New Method for Predicting Processability of Cellulose Compounds," *SPE Tech. Papers* **12** (3), 5 (1965).
33. E. P. Cizek, "Injection Molding of Polyphenylene Oxide," *SPE J.* **20** (12), 1295 (Dec. 1964).
34. E. Overgage and O. P. Phillips, "Successful Injection Molding of High-Density Polyethylene," *Brit. Plastics* **37** (3), 132 (1964).
35. W. E. Filbert and T. M. Roder, "Improved Surface of Acetal Moldings," *SPE J.* **20** (2), 149 (1964).
36. R. J. Kunze, "Painless Polycarbonate Processing," *Plastics Technol.* **10** (2), 32 (1964).
37. S. R. Cunliffe and J. G. Hawkins, "The Injection Molding of Nylon 11," *Brit. Plastics* **36** (12), 682 (1963).
38. F. J. Pokigo and R. E. Flodman, "Injection Molded Thin-Wall Impact Polystyrene Containers," *SPE J.* **19** (3), 289 (1963).
39. J. F. Moore, "Technical Aspects of Molding Acrylics," *Plastics Technol.* **7** (1), 54 (1961).
40. W. R. Schlich, "Progress in Injection Molding," *Plastics Technol.* **12** (5), 50 (May 1966).
41. M. W. Riley, "How to Buy Injection-Molding Machines," *Plastics Technol.* **11** (6), 2 (1965).

42. "Selection Guide to Plastics Machinery," *Plastics Technol.* (1964–65).
43. R. H. Hoehn, "Injection Molding," *Plastics Technol.* **6** (2), 66 (1960).
44. G. B. Thayer, "What a Plastic Engineer Should Look For When Planning to Buy an Injection Molding Machine," *Plastics Technol.* **4** (5), 439 (1958).
45. F. E. Schneider, "How to Buy Auxiliary Equipment for Injection Molding," *Plastics Technol.* **12** (7), 64 (1966).
46. "Dry Coloring for Injection Molding," *Plastics Technol.* **4** (5), 447 (1958).
47. A. Peiter, "Die Eigenspannunger," *Plastverarbeiter* **16** (11), 664 (1965).
48. *Kunststoffe Plastics* **13,** 158 (1966).
49. D. F. Hoffman, "Injection Molding of Thermosets," *SPE Tech. Papers* **12–13,** 70 (Dec. 1966).
50. E. A. Davis, "Ready to Mold Reinforced Thermosets," *Plastics* **31** (5), 563 (1966).
51. E. W. Vaill, "Injection and Extrusion of Thermosets," *SPE Tech. Papers* **12** (3), 6 (1966).
52. A. D. Sokolov and I. F. Kanavets, "Determination of Rheological Properties and Injection Moulding Conditions for Thermosetting Plastics," *Soviet Plastics* **1** (3), 22 (1966).
53. H. M. Gardner and M. G. Burges-Short, "Injection Molding of Thermosetting Plastics," *SPE Tech. Papers* **12** (3), 4, 1966.
54. A. D. Sokolov and I. F. Kanavets, "Influence of Moulding Conditions for Thermosetting Plastics of Strength Properties of Articles," *Soviet Plastics* **1** (1), 35 (1966).
55. L. J. Zukor, "Injection Molding of Thermosets," *SPE J.* **13** (10), 1057 (1963).
56. "New-Injection Molded Thermosets," *Mod. Plastics* **41** (10), 144 (1963).
57. C. D. Shaw, U.S. Pats. 2,296,295 and 2,296,296 (Sept. 22, 1943).
58. "Injection Molding Rubber Footwear," *Rubber Age* **68,** (1967).
59. "Injection Molding," *Rubber World* **148,** 29 (July 1963).
60. M. A. Wheelans, "Injection Moulding of Natural Rubber," *Rubber Developments* **18,** 133 (1965).
61. D. L. Daniels, "Economics of Rubber Injection Moulding—Single Station and Multistation Machines," *Rubber and Plastics* **46** (10), 1165 (1965).
62. M. A. Sanders, "Interrupted Cycle Molding," *SPE. J.* **17** (6), 557 (1961).
63. E. O. Prout, "Successful Injection Molding of Rigid PVC," *SPE J.* **23** (6), 75 (June 1966).
64. E. O. Prout, "Injection Molding of Rigid PVC," *SPE Tech. Papers* **11,** 16 (1965).
65. T. W. Moffitt, "Unplasticized PVC for Injection Moulding," *Plastics* **30** (11), 115 (1965).
66. H. H. Frimberger, "PVC Injection Molding Practices in Europe and U.S.," *SPE Tech. Papers* **11,** 6 (1965).
67. D. H. Jones, "Some Aspects of P.V.C. in Footwear Industry," *Rubber Plastics Age* **47** (2), 155 (1966).
68. "Progress in Footwear Moulding Equipment," *Rubber Plastics Age* **47** (2), 155 (1966).
69. W. E. Cohn, "Progress in PVC Molding in Footwear Industry," *SPE Tech. Papers* **11,** 28 (1965).
70. R. J. McCutcheon, "Molding of PVC Heals and Soles," *SPE Tech. Papers* **11,** 19 (1965).
71. A. R. Morse, "What Injection Molders Should Know About Reciprocating Screw-Tip Shut-offs," *Plastic Technol.* **13** (7), 46 (1967).
72. M. J. Hillier, "Analysis of the Inline Screw Injection Machine," *Polymer Eng. Sci.* **7** (3), 175 (1967).
73. P. N. Richardson and J. C. Houston, "Metering Screws in a Screw Injection Molding Machine," *SPE J.* **21** (1), 44 (1965).
74. J. Newlove, "How to Troubleshoot Your Screw Machine," *Plastics Technol.* **13** (5), 46 (1967).
75. V. R. Grundmann, "How to Evaluate Screw Injection Machine Plasticating Performance," *Mod. Plastics* **44** (11), 117 (1966).
76. D. I. Marshall and I. Klein, "Fundamentals of Plasticating Extrusion," *Polymer Eng. Sci.* **6** (3), 191 (1966).
77. Z. Tadmor, "Fundamentals of Plasticating Extrusion," *Polymer Eng. Sci.* **6** (3), 185 (1966).
78. B. Olmstead, "How to Start-up, Operate and Purge a Reciprocating Screw Machine," *Plastics Technol.* **12** (4), 37 (1966).
79. D. F. Moscher, "Why Screw Machines Over Plunger Machines," *SPE Tech. Papers* **12** (1), 5 (1966).
80. F. E. Schneider, "Quality Produced By Reciprocating Screws," *SPE Tech. Papers* **12** (1), 9 (1966).

81. M. Jury, "Screw Injection Machine Design," *Mod. Plastics* **42** (4), 119 (1965).
82. W. B. Campbell and P. H. Noble, "Shall I Convert My Plunger Machines?," *SPE J.* **20** (9), 1011 (1964).
83. W. C. Filbert, "Screw Plastication—A Status Report," *Mod. Plastics* **41** (7), 123 (1964).
84. M. L. Collins and R. H. Whitfield, "A Comparison of the Effects of Screw and Ram Injection Processes on the Physical Properties of Polystyrene Moldings," *Plastics* **29** (5), 65 (1964).
85. P. R. Schwaegerle and J. W. Poarch, "Optimizating the Operating Variables of a Screw Injection Molding Machine," *SPE J.* **20** (3), 262 (1964).
86. C. L. Weir, "Getting the Most Out of Screw Injection Molding Machines," *Plastics Technol.* **41** (3), 139 (1964).
87. P. T. Zimmermann, "Shot to Shot Variation on Screw Injection Molding Machine," *SPE. J.* **19** (10), 1061 (1963).
88. C. J. Waechter and L. J. Kovach, "Converting to Screw Plastication," *Mod. Plastics* **40** (3), 125 (1963).
89. C. L. Weir and P. T. Zimmerman, "Fact and Figures on Ram Versus Screw Injection, Part I," **40** (11), 123 (1962); "Part II," **40** (12), 125 (1962).
90. W. G. Kriner, "Graphic Comparison of Screw and Plunger Machine Performance," *Mod. Plastics* **39** (5), 121 (1962).
91. "Injection Machine," *Brit. Plastics* **39** (6), 343 (June 1966).
92. A. Haban, *Plasticke Hmoty a Kaucuk* **2,** 171 (1965).
93. J. C. Houston "Injection Molding," *SPE J.* **22** (6), 42 (June 1966).
94. G. T. Byast, "Viewpoints on Injection Molding," *Plastics Inst. Trans. J.* **32** (10), 302 (1964).
95. W. Laeis, "Precision Injection Moldings," *Plastics* **29** (3), 46 (1964); **29** (4), 59 (1964).
96. P. A. Plasse and A. D. Little, "Injection Molding," *Plastics Technol.* **9** (10), 50 (1963).
97. L. I. Johnson, "Strain-free Injection Molding," *Mod. Plastics* **40** (6), 111 (1963).
98. J. Newlove, "How to Troubleshoot Your Plunger Injection Machine," *Plastics Technol.* **12** (11), 39 (1966).
99. Z. Tadmor, "Non-Newtonian Tangential Flow in Cylindrical Annuli," *Polymer Eng. Sci.* **6** (3), 203 (1966).
100. I. Klein and D. I. Marshall, "Fundamentals of Plasticating Extrusion," *Polymer Eng. Sci.* **6** (3), 198 (1966).
101. F. E. Duffleid, "Melt Pressure Measurement and Control," *Plastics* **30** (5), 80 (1965).
102. J. F. Carley, "Fundamentals of Melt Rheology, Heat Generation, and Heat Transfer as Applied to Polymer Processing," *Polymer Eng. Sci.* **6** (2), (1966).
103. "Preplasticising Conversions," *Rubber Plastics Age* **46** (12), 1380 (1965).
104. H. W. Ashton and R. T. Cassidy, "Accurate Measurement of Nozzle Pressure During Injection Molding," *SPE J.* **19** (3), 295 (1963).
105. P. R. Junghaus, "Dispersion Aids for Dry Coloring in Injection Molding," *SPE J.* **18** (10), 1267 (1962).
106. C. L. Weir, "Multi-Gating Techniques for High-Density Polyethylene," *Plastics Technol.* **6** (5), 43 (1960).
107. T. P. Murphy, "Processing of Fiber Glass-Reinforced Thermoplastics Improved by Hot-Runner Molding," *Plastics Design Process* **12** (5), (1964).
108. A. Spaak, "Mold Design for High-Density Polyethylene," *Plastics Technol.* **4** (6), 537 (1958).
109. M. I. Ross, "Mold Design for Automatic Runnerless Molding," *SPE J.* **21** (6), 559 (1965).
110. F. J. Lupton, "Some Considerations on Combining Sprueless Molding Design Techniques with Auto-Unscrewing Molds," *J. Plastics Inst. Trans.* **34** (113), 261 (1966).
111. "Electroformed Textured Molds," *Plastics* **31,** 345, 906 (1966).
112. J. Jarrett, "Mold Design Related to Materials," *Brit. Plastics* **39** (5), 290 (1966).
113. W. E. Rowe, "Tool and Machine Requirements for Precision Molding," *SPE Tech. Papers* **12** (1), 12 (1966).
114. R. C. Perkins, "Cast Injection Molds for Quick Change Production Runs," *Plastics Eng.* **5** (9), 313 (1965).
115. "Presses Produce Larger Parts from Mold That Gives," *Iron Age* **196** (8), 84 (1965).
116. "Epoxy Molds—Easy Way to Get Injection Molded Prototypes," *Plastics Design Proc.* **5** (7), 12 (1965).
117. E. W. Vaill, "Essentials of Mold Design in Production Planning," *SPE J.* **21** (3), 274 (1965).

118. W. D. Campbell, "Mold Design for New Injection Machines," *SPE J.* **21** (1), 37 (1965).
119. P. L. Barrick, R. H. Crawford, and E. E. Sawin, "New Inexpensive Mold for Polymer Evaluation," *SPE J.* **20** (1), 69 (1964).
120. E. J. Csaszar, "Moldmaking and Tooling," *Plastics Technol.* **4** (7), 655 (1958).
121. R. N. Farris and R. J. Meeks, "Importance of Mold Rigidity in Injection Molding," *Plastics Technol.* **3** (5), 371 (1957).
122. E. Baer, ed., *Engineering Design for Plastics,* Reinhold Publishing Corp., New York, 1964, p. 455.
123. J. C. Engman, "Gates for Injection Molding," *Plastics Technol.* **8** (10), 23 (1962).
124. S. E. Giragosian, "Continuous Mold Venting," *Mod. Plastics* **44** (11), 122 (1966).
125. R. Sonntag, "Tolerances of the Finished Product as a Function of Control Possibilities of Machines," *Plastics* **30** (9), 87 (1965).
126. R. L. Peters, "Here's a Handy Way to Determine Plastics Shrinkage," *Plastics Technol.* **10** (9), 54 (1964).
127. H. Holt, "New Techniques in Shrinkage Control," *SPE J.* **20** (6), 519 (1964).
128. "Progress Report on Mold-Release Agents," *Plastics Design Proc.* **5** (3), 10 (1965).
129. "Two-Color Injection Machine," *Plastics* **31** (4), 434 (1966).
130. L. Temesvary, "Mold Cooling—Key to Fast Molding," *Mod. Plastics* **44** (12), 125 (1966).
131. C. E. Waters, "Thermoplastic Molding Its Cooling Needs," *Plastics Technol.* **11** (12), 25 (1965).
132. "Hansson Mould Temperature Controllers," *Intern. Plastics Eng.* **5** (11), 397 (1965).
133. "Efficiency Recorders in Plastics Molding," *Plastics* **30** (5), 75 (1965).
134. W. Bauer, "The Interdependence of Different Settings in Injection Molding Machines," *German Plastics* **55** (4), 12 (1965).
135. G. Stimpson, "Mold Temperature Control: A Study of the Factors Involved and Methods Used," *Intern. Plastics Eng.* **5** (3), 65 (1965).
136. "Heating Systems for Plastics Processing Machines," *Plastverarbeiter* **15** (1), 12 (1964).
137. K. J. Cleereman and D. S. Chisolm, "Copper-Clad Cylinders for Induction Heating," *Mod. Plastics* **39** (6), 131 (1962).
138. W. E. Lorton, "Induction Heating in Plastics Machinery," *Intern. Plastics Eng.* **4** (12), 348 (1964).

General References

E. Baer, ed., *Engineering Design for Plastics,* Reinhold Publishing Corp., New York, 1964.

J. T. Bergen, "Mixing and Dispersing Processes," in E. C. Bernhardt, ed., *Processing of Thermoplastic Materials,* Reinhold Publishing Corp., New York, 1959, Chap. 7, p. 405.

E. C. Bernhardt et al., *Processing of Thermoplastic Materials,* Reinhold Publishing Corp., New York, 1959.

J. Delmonte, *Plastics Molding,* John Wiley & Sons, Inc., New York, 1952, pp. 33–156.

S. Levy, "Injection Molding" in J. Frados, ed., *Modern Plastics Encyclopedia,* McGraw-Hill Publishing Co., 1966, Vol. 44 (1A), p. 710.

W. Mink and E. G. Fisher, *Practical Injection Moulding of Plastics,* Iliffe Books, Ltd., London, 1964.

E. T. Severs, *Rheology of Polymers,* Reinhold Publishing Corp., New York, 1962.

G. B. Thayer, J. W. Mighton, R. B. Dahl, and C. E. Beyer, "Injection Molding," in E. C. Bernhardt, ed., *Processing of Thermoplastic Materials,* Reinhold Publishing Corp., New York, 1959, Chap. 5, p. 308.

I. Thomas, *Injection Molding of Plastics,* Reinhold Publishing Corp., New York, 1947, pp. 365, 389.

S. E. Tinkham, "Injection Molds for Thermoplastics," in J. H. DuBois and W. I. Pribble, eds., *Plastics Mold Engineering,* Reinhold Publishing Corp., New York, 1965, Chap. 9, p. 397.

J. S. Walker and F. R. Martin, *Injection Molding of Plastics,* Transatlantic Arts, Inc., New York, for the Plastics Institute, London, 1967.

"Injection Molding" in *Modern Plastics Encyclopedia,* McGraw-Hill Book Co., 1966, Vol. 43 (1A), pp. 712–753.

Plastics Engineering Handbook of the Society of the Plastics Industry, 3rd ed., Reinhold Publishing Corp., New York, pp. 406–495.

Lee J. Zukor
Allied Testing and Research

BLOW MOLDING

Blow molding is used for producing bottles or other hollow objects from thermoplastic materials. In the modern process for blow molding of plastics a "parison" or tube is introduced into a mold in a molten or softened condition and expanded by the application of internal pressure against the walls of a mold, where the plastic hardens as it cools. The parison can be a hot tube extruded and placed immediately into the mold, a vial-like tube produced in an injection mold and transferred immediately to a separate blow mold, or a length of cold plastic tubing that is subsequently reheated prior to insertion into a mold for blowing.

Practically any thermoplastic material can be blow molded. Both branched and linear polyethylene as well as rigid formulations of poly(vinyl chloride) are used in quantity in the manufacture of containers, toys, and household and industrial products by blow molding. Polycarbonates, methacrylate polymers, acetal resins, and polystyrene are among the polymers used in blow-molded specialty products. The blow molding of plastics is continuing to expand at a rate faster than that of the plastics industry as a whole.

History

The first known use of a blowing process to produce objects dates back to ancient times when animal horns and tusks were softened by solvents and heat and blown into simple hollow vessels. This was followed by the blowing of a lump of molten glass on the end of a hollow reed into vessels of intricate design. Although there is a strong similarity between glass and plastic bottlemaking, the term blow molding appears to have come into use with the advent of plastics. Glass-bottle manufacture has been and still is referred to as "glass blowing" in the industry.

The first commercial blow molding of plastic consisted of blowing two sheets of cellulose nitrate between two closed mold halves which sealed the periphery of the item; introducing steam or air between the sheets stretched the hot sheet to meet the contour of the molds. However, owing to the highly flammable nature of cellulose nitrate the items fabricated from it had limited usefulness.

Early in the 1930s with the advent of injection molding and the less flammable cellulose acetate plastic molding materials, attempts were made to automate blow molding of plastics by combining some of the basic principles of injection molding and extrusion with some of the mechanics of the glass-blowing industry. With the commercial availability of low-density polyethylene following World War II came the production of plastic bottles by blow molding, and during the 1950s a wide variety of blow-molded products based on low-density polyethylene were introduced on the market. These included bottles for cosmetics and toiletries, pharmaceuticals, and chemicals, dispenser containers for mustard and ketchup, as well as an occasional industrial product such as toilet floats. For large-volume production, however, blow molding had to await the commercial development of high-density polyethylene. Early in the 1960s, an unprecedented expansion of blow-molding facilities took place, accompanied by general public acceptance of plastic bottles.

Characteristics of Blow Molding

There are a number of technical and economic advantages that the blow-molding process offers in the manufacture of plastic items. These advantages are: the possibility of reentrant curves, low stresses, the possibility of variable wall thickness, use of polymers with high chemical resistance, and favorable cost factors.

Reentrant curves are without doubt the most prominent feature in blow-molded items; this is true to the extent that it is difficult to find examples that do not incorporate them. No doubt this is due to the fact that such curves are included in the esthetic design for merchandising reasons even when they are not absolutely necessary for the functional design of a particular product. The word most commonly used for this type of construction is "hollow," but Webster defines hollow as "any concavity like a bowl or a drinking glass." Such hollow shapes are injection molded in plastic on a large scale. When walls are designed so that they turn inward toward the center, they are "reentrant." Blow molding has become the most prominent method for manufacture of plastic items with re-entrant curves.

In comparing injection molding with blow molding, a very significant technical difference is obvious. Blow molding is accomplished at pressures of from 25 to 100 psi between the plastic and the mold surface. Injection molding requires pressures of from 10,000 to 20,000 psi, at which the plastic is pushed through long, narrow passages; these high pressures result in orientation of the polymer and its resulting disproportionate stress distribution (see also the section of this article on Injection Molding). The lower internal stresses in the blow-molded items result in improved resistance to all types of strain (tensile, impact, bending, environmental, etc) and consequently a product with superior operating properties.

Since the mold equipment required for the blow-molding process consists of female molds only, it is possible, by simple changes of machine parts or extrusion conditions at the parison nozzle, to vary the wall thickness and thus the weight of the finished product. For an item where the exact thickness required in the finished product cannot be calculated with accuracy in advance, this is a great advantage in both time and costs. In injection molding, making such changes for each variable would be expensive. In vacuum forming, such changes would require availability of an infinite variety of sheet thicknesses. See also BAG MOLDING.

With blow molding it is possible to produce walls that are almost paper thin. Such thicknesses cannot be produced by injection molding at all, but they *can* be produced by vacuum forming. Both blow molding and injection molding can be successfully used for very thick walls. The final choice of process for a specific wall section would be strongly influenced by other factors such as re-entrant curves, stress, etc.

Blow molding can be used with some plastics, such as polyethylene, having a much higher molecular weight than can injection molding. For this reason, items can be blow molded utilizing the higher permeability, oxidation resistance, ultraviolet resistance, etc, of the higher-molecular-weight plastics. This feature is very prominent in resistance to environmental stress cracking. This extra resistance is necessary for plastic bottles used for detergents and for packages and parts in contact with many industrial chemicals that promote stress cracking. See also Long-Term Phenomena under FRACTURE.

The element of cost is so complex that it is difficult to generalize. The cost factors of blow molding are very pronounced in the area of capital equipment portion of the cost. Pound for pound of finished product, blow-molding-machine costs average between one-half and one-third of injection-machine costs for the same or similar items. Molds for blow molding have been known to cost as little as 10% of the cost of injection molds for comparable items.

These cost advantages do not necessarily apply to consumer container processing where the high unit volumes and the degree of automation have increased the costs of both machines and molds into ranges comparable to those of injection molding. At the same time, however, the increased efficiency of automated systems has brought unit costs of blow-molded items down to levels comparable with those of similar injection-molded items.

Molds

In its simplest form, a blow mold consists of two female cavities which close around a tube or other hollow form of hot plastic (see also MOLDS). Pinch-offs at one or both ends are used to define the extremities of the item. An entrance into the part is required for the blowing air and some means of cooling is usually provided. A cut-out section is required in the bottom (or top) of the die to accommodate the excess tube.

To the inexperienced eye, molds for blow molding may not look particularly different from conventional molds for injection molding. The basic differences are few in number, but they far outweigh the similarities. These differences are: (a) blow molds use cavities only; (b) blow molds do not require hardening, except for long production runs; (c) material movement within a blow mold is a stretching rather than a flowing action; and (d) undercuts on hollow blow-molded items, particularly polyethylene, can be stripped from the mold. The first difference, use of cavities only, eliminates at the mold-construction stage the complexities of matching forces with cavities as well as the problems of wall thickness usually associated with it. The wall-thickness problem is not eliminated entirely from the overall problem; the bulk of this problem is merely shifted from the moldmaking to the design and processing stages.

The use of unhardened cavities permits the use of many lower-cost procedures with a resulting lower overall cost for the completed mold. Unhardened cavities are feasible in blow molding because of the very low internal molding pressures required. Hardening, where required in blow molds, is usually limited to the use of hardened inserts.

Molding by a stretching rather than a flowing action reduces but does not eliminate the problems of weld and flow lines, mold erosion, trapped gas, etc. The problems of blow molding more closely parallel those of vacuum or thermoforming than either injection or compression molding. In turn, the construction details of blow molds show many similarities to those of molds used for thermoforming. See also BAG MOLDING; THERMOFORMING.

The possibility of stripping certain amounts of undercuts from the mold, due to the fact that the inner substance is air rather than steel, eliminates the need for expensive sliding inserts, levers, cams, etc, used extensively in injection molding for undercuts. Sliding inserts are used in blow molding to create completely enclosed shapes such as toilet floats, bowling balls, Christmas tree ornaments, etc, but not for objects with body undercuts.

Construction. Except for the differences discussed above, the design and manufacture of blow molds follow most of the same basic concepts that apply to conventional molds. As with conventional molds, blow molds can be produced from many different metals using more or less conventional metal processing equipment. The usual problems of cooling, machining and assembly, and weight and durability of the finished mold are as common here as in conventional molds. These are discussed in detail in the article MOLDS. The construction methods and materials commonly used for commercial blow molds are shown in Table 1. Metal spraying and electrolytic deposition methods have been used to a limited extent.

Table 1. Materials and Processing Methods Used for Manufacture of Blow Molds[a]

Material	Cooling preference[b]	Weight preference[b]	Turn	Duplicate	Hob	Pour cast	Pressure cast
beryllium–copper alloy	1	6				×	×
aluminum, cast	2	1				×	×
aluminum, aircraft	2	2	×	×			
zinc (Kirksite A)	4	3				×	×
iron	5	4				×	
steel	6	5	×	×	×		

[a] × indicates suitability of method to material listed.
[b] Numbered 1 to 6 in order of desirability.

Atmospheric and pressure casting are probably the most popular processes used for blow molds for medium-volume production runs. Technically speaking, almost any shape can be produced by the casting process. The resulting mold, however, is not always the most economical, versatile, or durable. Atmospheric casting has the advantage of low equipment cost, but the disadvantages of porosity and lack of toughness in the material, which may make insert work difficult. Appearance details such as surface texture and engraving are more uniform from cavity to cavity than can be achieved by engraving on machined cavities. Atmospheric casting has limitations as to the details that can be cast, due to trapped air. This has led to the development of pressure-cast cavities.

Pressure casting is superior to atmospheric casting in reproducibility of fine details, sharp angles, and straight sides. Zero-draft side walls and engraving as small as typewriter type have been successfully reproduced by this process. It also produces a tougher, less porous metal allowing the addition of inserts and machining after casting. It has the disadvantage of a high initial hob cost and, at the present time, limited availability.

Turning is excellent for a small number of small round dies. It permits the economical use of steel, which, with its inherent toughness, allows the use of interchangeable inserts for maximum design flexibility. Duplicating is a reliable process for blow molds. Its availability and predictability are offset, however, by its higher cost, which is a serious deterrent for small- to medium-volume production. For high-volume production where toughness and long life are required, combined with maximum cooling capacity, blow molds duplicated from bar-stock aircraft aluminum have proved to be well worth the extra cost. Press hobbing has generally been too expensive a process for blow molds. It would be of value mainly where hardened steel is desired in the entire mold; such a condition might arise on long runs with multiple cavities subject to hard usage. Processes having extremely fast mold closing or closing under pressures in the range of injection molding, as used in the injection-blow process, can justify the added expense of hardening. In some types of blowing the entire circumference is pinched off by the die, and here hardened dies can sometimes be justified.

If an item is basically round and smooth in cross section, it can usually be turned on a lathe. Template turning is needed for return curves. For turned molds, inserts in either or both top and bottom are required in order to gain entrance for the tool in turning. Inserts, both top and bottom, are advisable wherever close tolerances, such as for snap fits, threads, and push-up bottoms, are required or when interchangeability is desired.

When multiple-cavity molds with no inserts are involved, casting can sometimes be cheaper than turning. Multiple-cavity round molds favor casting, particularly if the molds are side by side and a one-piece construction is feasible. One-piece construction avoids the problem of heat loss at the interfaces of insert. Variations can be produced in casting by incorporating inserts in the original hobs and making separate molds for each variable. Any shape that cannot be turned is considered irregular and is usually duplicated or cast.

Choice of the metal to be used for the mold is influenced by specific gravity (determining weight of the mold), toughness, and rate of heat transfer. Since rate of cooling has a large degree of control over overall production cycles, the heat-transfer properties of the mold are of considerable importance to the economics of blow molding.

Design. Design of a blow mold requires consideration of the following factors: material to be blown; size and weight of product and mold; contour; variations of design desired, eg, neck inserts, bottom plates, or side inserts; surface texture and engraving; sharp corners and straight angles; blow openings available and location; and parting lines. Of the multitude of factors affecting the design or construction of blow molds, the most influential is the design of the item to be blown. Exaggerated size relationships, such as extremely narrow or extremely wide tops, bottoms, or midpoints, are not only difficult to blow, but the difficulty is frequently reflected and sometimes magnified in the mold construction. The material is also of prime consideration. For example, low-density polyethylene is soft and can often be pinched off using cast aluminum pinch-offs, whereas linear polyethylene is tougher. Steel inserts are popular. For use with linear polyethylene, particularly in soft metal molds, pinch-offs in aircraft aluminum and beryllium–copper molds have been successfully machined directly into the metal.

When a product design is confined to a single style with no variations, the mold-making problem is greatly simplified and strongly favors casting. In many cases, however, it is desired to have a variety of style changes incorporated in a single die. A classic example of variety of styles in a single mold is a cylinder bottle with interchangeable neck inserts and bottom plates affixed to each end of a straight cylinder. In this case, machining is usually the most economical method of construction, using steel throughout to resist the wear from repeated changes of inserts. For irregular shapes, duplicating in steel would be advisable, although atmospheric casting of iron or pressure castings in zinc or beryllium–copper alloy have also been used successfully in this type of construction. Surface textures and engraving strongly favor pressure casting.

Blow Opening. The nature of the blow opening available is in itself of no definite consequence in the decision of what method to use to make a die, but the details of construction may be greatly influenced by this factor. A closed blown part will require sliding pinch-off at the blow point to seal off the finished part. Parts with only a small hole require needle blowing. Needles can be introduced practically anywhere on a part. Additional problems arise, however, if, for instance, the needle is desired in an area adjacent to water lines or cooling areas, or at the top or bottom of the part, requiring movement of the needle during the cycle. The type of molding machine available has a strong influence upon the blow opening used. Some machines, for instance, use needle blowing exclusively, excess material being machined off to provide larger openings when required.

Parting Lines. Single-plane, vertical parting lines are customary for most blow molds. In some cases, the product is blown at an angle other than its normal mid-axis. This requires tipping the cavities and can be accomodated by all methods of construction. In some cases, as with a figurine, it is necessary to have a parting line that is not in a single plane. Such cavities would be best cast in steel or beryllium–copper. A close miter can be made economically and the steel or beryllium–copper will give strength to the parting line to resist the wear this type of mold is subjected to in operation.

Tolerances and Fit. Close tolerances on the capacity of the blown item also influence die construction. Where multiple molds are required, capacity from one mold to another must be closely held. In general, some of the casting processes have been more successful in reducing this factor than duplicating or hobbing. For final capacity adjustment, an inserted bottom plate or other adjustable insert is advisable. By this method, slight changes can be made on each mold insert without noticeable variation in finished parts.

Plastic will sometimes thin out at an insert or parting line. With some materials, like low-density polyethylene, this is of little importance, but with high-density polyethylene and similar materials it can be a serious problem. When inserts are used they should be fitted tightly. Parting-line miter should be as close as possible and sufficient clamping pressure should be provided to prevent partial opening of the dies. Solid molds have a slight technical advantage over molds with inserts in these instances.

An interchangeable fit with other parts requires serious consideration in mold construction. This type of construction includes plug fits, screw- and snap-type caps, snap beads, doll parts, etc. It is strongly advisable to use inserts for the interchangeable portion. This would allow for changes of inserts to accommodate different shrinkage conditions, different materials, etc.

Surface. The surface required also influences mold construction. Surface textures and engraving strongly indicate pressure casting.

A high surface luster is usually desired for parts requiring detailed printing. Some blown polymers, such as poly(vinyl chloride), produce a luster from a highly polished mold surface, others, such as polyethylene, require sand blasting or other roughening of the surface to produce luster.

Polytetrafluoroethylene and various other coatings have been successfully used on molds to produce luster. Plating has been used but does not seem to be as successful as it is in injection molding. Cast-iron molds, however, should definitely be plated for release, surface luster, and resistance to rusting. Silicone release agents will cause subsequent peeling if the product is to be printed or painted.

The mold temperature will affect the quality of the surface of the blow-molded product. Surface quality is also affected by blowing pressure, material temperature, and the presence of trapped air or moisture in the mold cavity.

Venting. Roughening of the surface permits the release of trapped air. A mold with several inserts (Fig. 1) will often release air through the interfaces of the inserts without extra venting. In severe cases of air entrapment, such as in deep engraving, vent holes of 0.004–0.007 in. are necessary. Conflict of vent holes with cooling channels is a common problem.

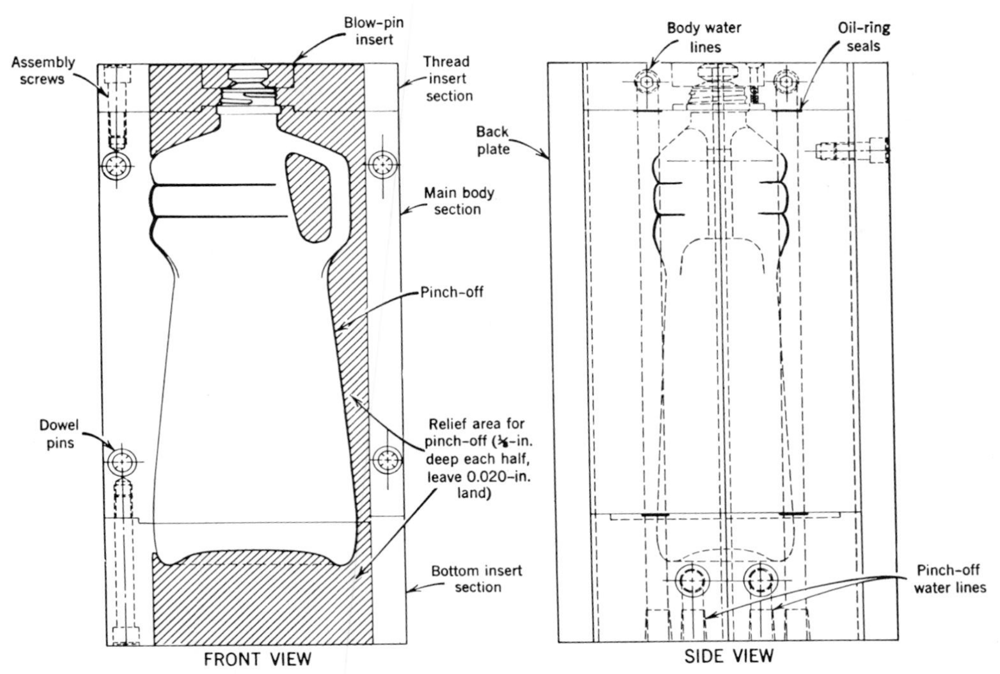

Fig. 1. A three-section blow mold made by machine duplicating from densified aircraft aluminum. Sectional construction is used to permit changes in bottle design. In this mold the thread size can be changed or new base added to produce a longer bottle. The interfaces of the inserts have the added advantage of providing good venting conditions.

The blow air expands the parison rapidly, and the mold must be properly vented to allow the air in the mold to escape quickly. Trapped air will cause surface defects on the object at points where there are sharp radii, joints between mold inserts, and the parting lines. Surface depressions are a problem, since displaced air becomes trapped unless proper venting is accomplished.

Temperature Control. Temperature control is necessary in blow molding. Cooling channels should be installed as close as possible to the mold cavity, and should be baffled to increase the surface area with which cooling water is in contact. A low-pressure (30 psi) coolant supply system is common. Higher-pressure systems are difficult to maintain owing to the possibility of leakage through porous portions of the mold metal or through joints in segmented molds.

If a soft mold material is used, steel inserts should be used for the neck; both the bottom pinch-off (Fig. 2) and the neck insert should be separately cored for cooling because of the greater amount of heat that must be removed owing to the greater thickness of the material.

To obtain good quality with maximum stress resistance, refrigerated coolant should be passed through the neck ring and bottom pinch-off. Coolant in the range of 35–70°F should be used for the mold body. Too low a temperature in the pinch-off area can have a detrimental effect on the blowing of the parison tube. When the pinch-off plates squeeze the parison, the material in contact with the cold surface of the pinch-off crystallizes and cannot blow outward to the extremities of the molding. This causes thin walls at the extremities. Too hot a pinch-off is also detrimental,

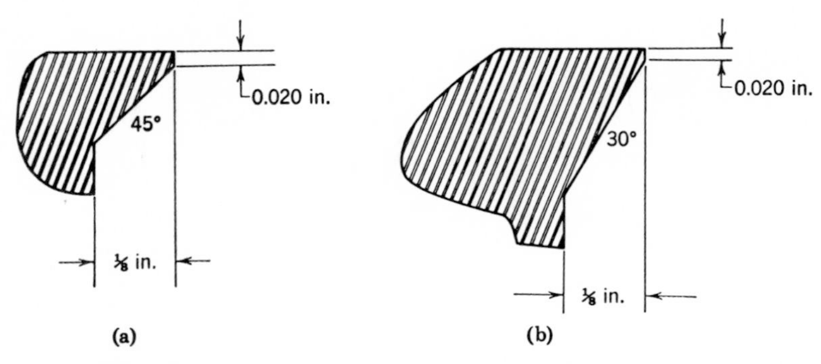

Fig. 2. Pinch-off design for use with linear polyethylene and most other thermoplastic materials. Parison wall thickness for all designs, 0.150 in. Only one side of the mechanism is shown; compare with Figure 3. (**a**) Typical pinch-off design for handles and circumferential pinch-off areas. (**b**) Typical pinch-off design for bottom pinch. Design as shown is for a flat area adjacent to pinch-off.

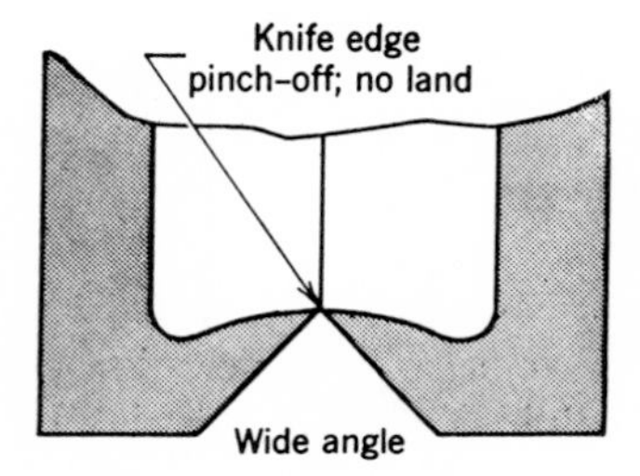

Fig. 3. Unacceptable pinch-off design. Knife edge is too sharp and the wide angle prevents holding and cooling of excess plastic. This design will suffice for certain situations, particularly low-tolerance products in low-density polyethylene.

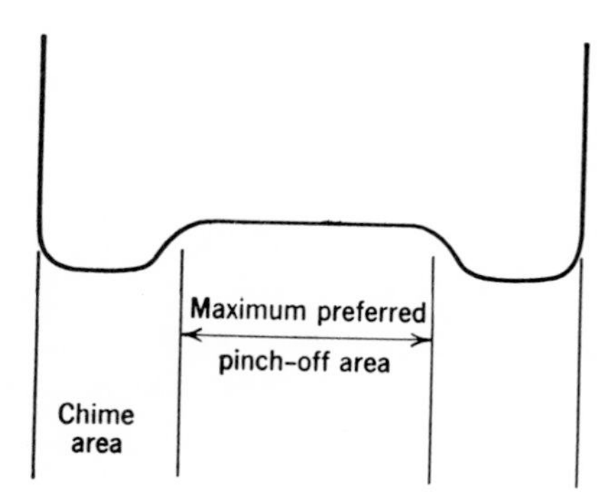

Fig. 4. Diagram showing chime and area where pinch-off should be located.

since it produces the opposite effect (thin sections at the pinch-off and heavier walls at the extremity of the product).

Cold mold temperatures require high blowing pressures to produce good surface finishes. The exact temperatures for each section of a blow mold depends upon the product design. Properly balanced zone cooling will result in faster cycles with lower levels of strain in the finished product. Mold temperature and temperature control must also take into consideration the properties of the plastic being blown, required properties of the product, and shrinkage of the molded part. In general, the lower the mold temperature, the less shrinkage takes place, other conditions being equal.

Pinch-Off. A badly designed pinch-off can lead to a poor weld and excessive thinning at the weld. In extreme cases containers may be incompletely blown due to the formation of a hole in the pinch-off area. The pinch-off must also be designed to permit easy removal of flash.

Figures 2 and 3 show good and bad pinch-off designs for linear polyethylene. The same or similar designs have been found to be satisfactory with poly(vinyl chloride), polystyrene, and other common thermoplastic materials. The exact width of the land may require alteration to meet a specific molding condition or product design. For example, tough engineering-type materials such as the acetal resins and polycarbonates require much smaller pinch-off widths. Pinch-off for these materials should be in the range of 0.002–0.005 in. The relief area should be shallow enough to cool the plastic, but if it is too shallow it may prevent the mold from closing completely. A thickness of about 80–90% of the total thickness of plastic in the parison tube has been found to be most satisfactory.

If possible, pinch-off for bottles should be designed so as not to extend into the chime of base (Fig. 4) to avoid weakening the chime against drop impact and environmental stress-cracking conditions.

Extrusion Head and Die

All blow-molding machines, with the exception of the injection-blow type (see section on Injection Blow Molding) and those which blow an object from a preform, have an extrusion head from which the parison is extruded. The design of this head is very important in determining the quality of the parison, and hence the product.

Flow Control. The inside surfaces of the flow channels within the head, including the die and the core, should be well polished to prevent any accumulation of the plastic flowing over these surfaces. If the surface of the channel is rough, stagnant layers, or "hang-up," of the polymer may form. This material will eventually degrade, causing dark streaks to appear in the parison. Roughness in flow channels may also cause the parison to rupture or blow out in affected areas during its expansion in the closed mold.

Streamlining the shape of the flow channels also helps to prevent areas of stagnation and aids in reuniting the plastic after it has flowed around both sides of the mandrel in the head (Fig. 5). At the point where the material rejoins after flowing around the mandrel, a small pocket of stagnant material can form if the flow channel in the head has been improperly streamlined. This stagnant pocket of material will then continually bleed degraded material along the weld line of the parison.

Many variations of streamlining are used. A common method of streamlining is to cut away the side wall of the mandrel at the point of impact with the incoming plastic melt. In some designs the point of contact is reduced to almost a knife edge. The cutaway portion of the mandrel is usually shaped into channels that flow downward as they circle the mandrel. At a point 180° from the entrance port and a little below the port, the two streams of plastic melt from each side of the mandrel are brought together. The metal of the cutaway mandrel at the point of confluence is also machined almost to a knife edge to minimize stagnant areas. In passing around a mandrel the pressure of the plastic melt will decrease. To compensate for this pressure drop, the cross section of the melt channel on the reverse side is usually reduced at the point of confluence as compared to the cross section 180° around the mandrel. The lower cross section increases the pressure on the melt, compensating for the loss and aiding the reuniting of the two streams of plastic into a homogeneous melt.

Healing of the weld line in the flowing plastic depends on three factors: retention time in the head, head temperature, and back pressure on the melt. Time is an important factor. Sufficient retention time must be provided to allow the hot plastic to intermix and cohere. Increases in head pressure and temperature will assist this process, but are less effective than an increase in time.

To increase head pressure, an annular restriction called a "choke" is often placed on the mandrel in the extrusion head downstream from the flow deflector. This dam decreases the cross-sectional area of the annular flow channel in the head and builds up melt back pressure in the weld area just past the flow deflector and upstream from the choke. The use of chokes and increased head pressure without a flow deflector is not effective in producing optimum weld conditions. If a choke is used and flow is not streamlined with a deflector, small, hard, carbonized particles of degraded polymer or hard gels become trapped in the channel.

Flow behavior of the melt is affected by the type of polymer being extruded. For example, linear polyethylene, a long-chain molecule with few branches (and the polymer most commonly used in blow molding), is more sensitive to changes in direction of flow than is branched, low-density polyethylene. Other polymers also vary in their relationship between flow behavior and shear rate.

Differentials in the velocity of a plastic melt are caused by friction of the plastic on the extruder surfaces and to some extent by internal friction within the plastic mass. Frictional drag is directly proportionate to the distance traveled. Pressure drop in the head of a blow-molding machine caused by changes in cross section is particularly acute at the right-angle bend in going from a horizontal extruder into a vertical parison die. Pressure differentials between the incoming side of the head and the opposite side of the core pin can be as much as 300–400 psi in a symmetrical head design. This pressure differential causes changes in the shear rate and velocity within the plastic mass. The highest velocity and shear rate occur directly below the inlet channel from the extruder; maximum internal pressure also occurs at this point. Minimum shear rates and velocities occur at a point 180° opposite on the other side of the core pin. Pressures between these two points drop more or less uniformly in a concentric head design. See also Melt extrusion.

A choke must be added slightly below the inlet channel to create a high pressure and to slow the flow on the inlet side of the core pin. The design and positioning of a choke to produce uniform flow is dependent on factors such as overall machine design, the output rate of the extruder, and the back pressure. A properly designed choke will compensate for the various differences in pressures and shear rates. The choke must not create turbulence in the flow pattern. Since there are no theoretical or empirical formulas which accurately predict the melt flow in the head, the actual flow patterns in a head design must be determined experimentally.

In addition to uniform flow conditions in the extrusion head, it is necessary to control the flow leading to the die orifice. A tapered approach angle to the die orifice is recommended for most applications. Each type of material will require a different approach angle. The ratio of wall thickness at the die orifice to the length of die land is a controversial subject. For linear polyethylene, for example, die lengths as high as twelve to twenty times the wall thickness are recommended by some material sup-

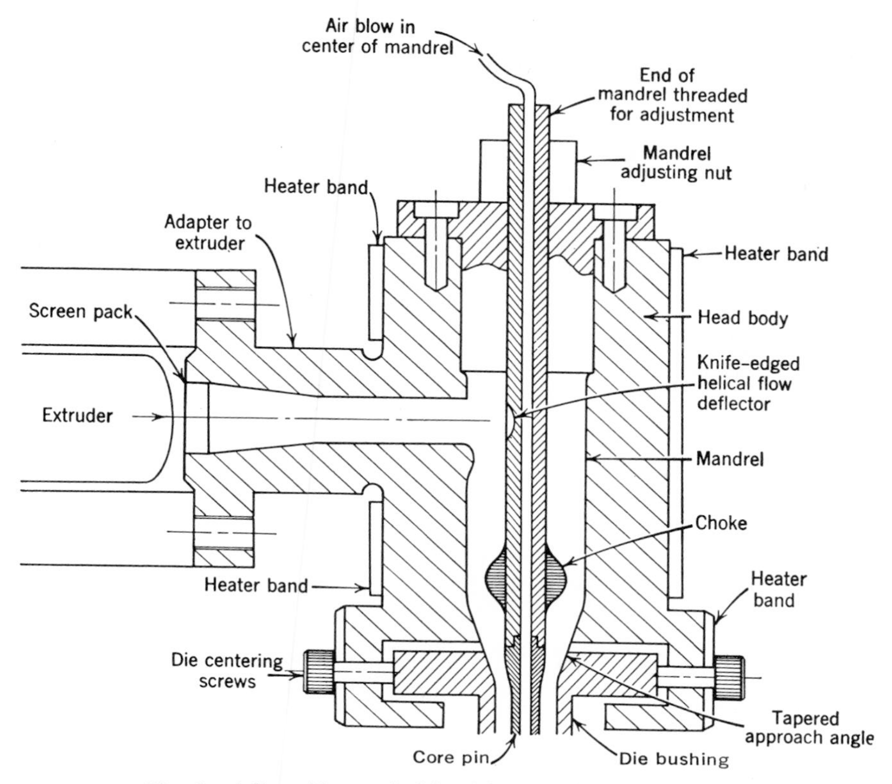

Fig. 5. Adjustable mandrel head for use with polyethylene.

pliers, whereas some blow-molding machines have operated successfully with linear polyethylene using minimal die lands approaching knife edges in thickness. Die-land specifications can, therefore, best be classified as dependent upon shop practice.

Die and Core-Pin Design. The thickness of the parison wall must be as uniform as possible in both the transverse and the longitudinal cross sections. The die and core pin (Figs. 5 and 6) determine the inside and outside diameters of the parison tube. The requirements for these dimensions are dictated by the shape and wall thickness of the item being produced. The die and core pin must be concentric to produce a uniform wall thickness. However, concentricity is not a guarantee of uniform wall thickness since imbalances in the plastic melt, such as pressure or temperature, can cause a nonuniform wall in a parison tube from a perfectly concentric die and core pin. Parison control techniques which alter the wall thickness of the parison in both the longitudinal and the cross-sectional directions are even more sensitive to unbalanced melt flow than are concentric parisons.

There are several variables which determine the dimensions of the die and core pin. The two most important variables are "blow-up" ratio and swell. Blow-up ratio is the ratio of the maximum outside cross-sectional dimension of the finished molded part to the maximum outside cross-sectional dimension of the parison. Normally, it is desirable to work with blow-up ratios in the range of 2–3 to 1, but it is possible for blow-up ratios to go as high as 4 or 5 to 1 in special thin-wall products. Die swell is the percent increase of the relaxed diameter of the parison over the diameter of the extrusion die. Die swell is affected by temperature and pressure of the melt, parison cross section, type of material, parison extrusion rate, die angle, and die/land ratio (see also MELT EXTRUSION). One of the most critical requirements for accurate control of blow-up ratio and swell is in the blow molding of plastic bottles for which the neck finish can have no flash or pinch-off. Under these conditions it is necessary to produce a parison that is only slightly smaller than the diameter of the neck finish after all swell has taken place in the parison.

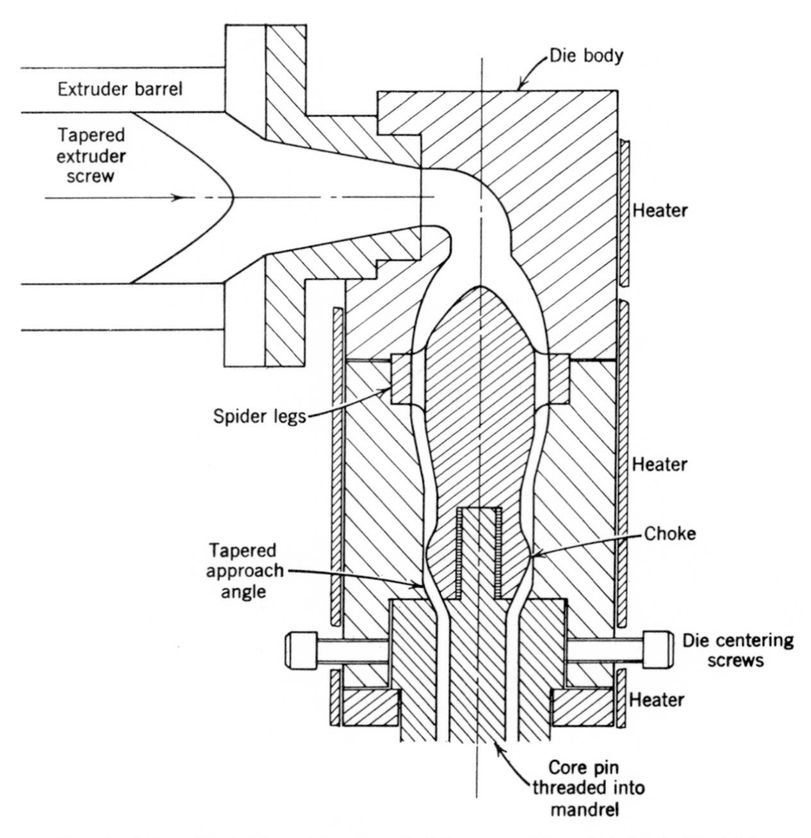

Fig. 6. Nonadjustable spider mandrel for use with poly(vinyl chloride).

Right-angled extrusion heads with core pins which work satisfactorily with polyethylene usually produce weld lines with poly(vinyl chloride). This is overcome by using a centrally fed spider head for poly(vinyl chloride), which minimizes the interruption of material flow and prevents movement of the mandrel under the high pressure generated.

The extruder employed should have a length/diameter ratio of 24:1 or greater and the screw a compression ratio of 3:1 or greater obtained by constant pitch, decreasing depth of flight. The possibility of melt "hang-up" with resultant decomposition is considerably less with a centrally fed spider than with a right-angled extrusion head. Figures 5 and 6 illustrate core-pin head designs for polyethylene and spider designs for poly(vinyl chloride).

The Parison. Control of parison thickness is a vital portion of all blow-molding processes. Parison control can be either mechanical or hydraulic; continuous extrusion processes cannot use hydraulic control and must accomplish parison control strictly by mechanical means. Accumulator systems, whether external or reciprocating-screw type, can use a change in hydraulic pressure on the ram to produce a change in thickness of the parison due to change in die swell. The hydraulic system works very rapidly but does not produce large changes in wall thickness; these must be accomplished by mechanical means, as illustrated in Figure 7. An ideal combination for parison control is an accumulator system with hydraulic variations and a mechanical parison control unit as well. This gives maximum versatility in the amount of wall thickness control. This system is widely used for manufacture of carboys and other large items. Parison-control heads can be substituted for fixed parison systems without radical changes elsewhere.

Finished necks can be produced on bottles during the molding cycle by the use of mandrel calibration. In this process a tube larger in diameter than the desired bottle neck is extruded down over a specially designed blowing mandrel. After the mold is closed and the bottle is initially blown against the mold walls, the mandrel is forced further into the neck of the bottle forming both internal and external dimensions of the neck in conformity to the machined shape of the driven pin (Fig. 8). To speed

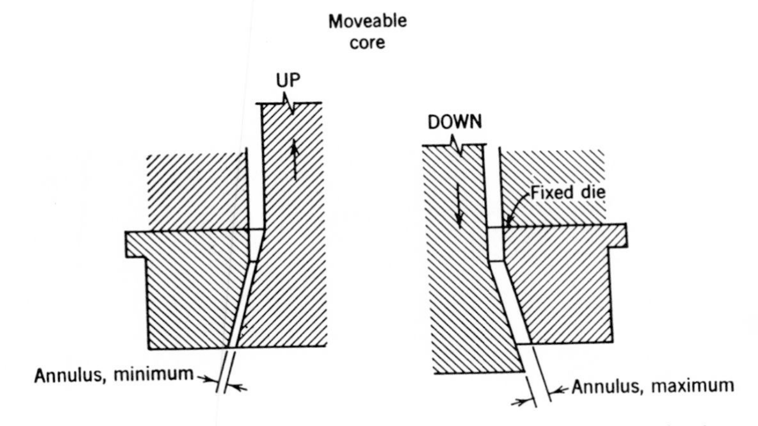

Fig. 7. Core-pin movement in a mechanical parison-control mechanism.

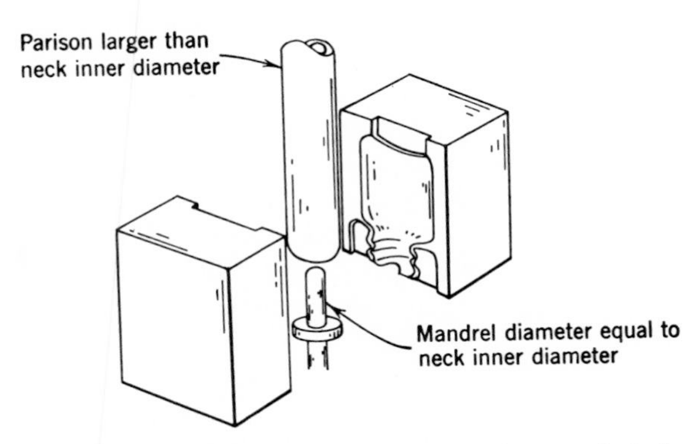

Fig. 8. Method of producing finished neck in bottles using mandrel of calibrated size.

the molding cycle the calibrating pin is usually cooled to give cooling on both sides of the plastic in the neck finish.

Extrusion Blow Molding

Manual Transfer. Manual transfer is still used in laboratories for sampling and testing, as well as in some specialty production. An operator seizes the top of a continuously extruded tube with tongs when it reaches the desired length, breaks away a portion of the tube, and conveys the tube to the blowing station where he inserts it between the mold halves and activates the closing of the platens. The blowing cycle is usually governed by a timer, but in this crude arrangement ejection of the finished part usually requires manual removal.

Intermittent Extrusion In the intermittent extrusion process the parison is extruded intermittently by stopping and starting the extruder. A horizontally opening stationary mold is mounted directly below the extruder cross head, and a blow pin or mandrel is either mounted from below the mold or through the center of the parison die in the cross head (Fig. 9). Blowing and ejection can be automated, but since this arrangement is usually for laboratory and test purposes, these functions are controlled by the operator and varied from cycle to cycle.

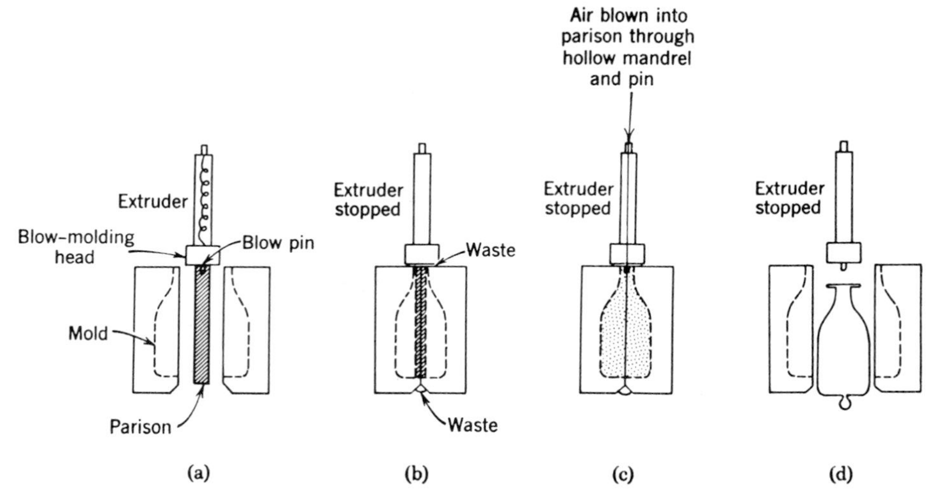

Fig. 9. Intermittent extrusion. (**a**) Complete extrusion of parison; (**b**) close mold; (**c**) blow and cool; (**d**) eject.

With intermittent extrusion, the full capacity of the extruder is not utilized. Cycle time and direct labor costs are high; however, the process may be practical for short sample runs and development work. Stop–start machines are not offered commercially as stock items to any extent, but there are many custom-built and homemade varieties in operation in production plants. This type of construction has the advantage of low capital investment and relatively few parts. Such equipment can be altered to suit the needs of the program.

Manifold Extrusion. An immediate offshoot of the simple laboratory machine is a machine for the continuous extrusion, and intermittent, valved-delivery blow molding. Such a machine delivers plastic from the extruder via manifolds (pipes) to sepa-

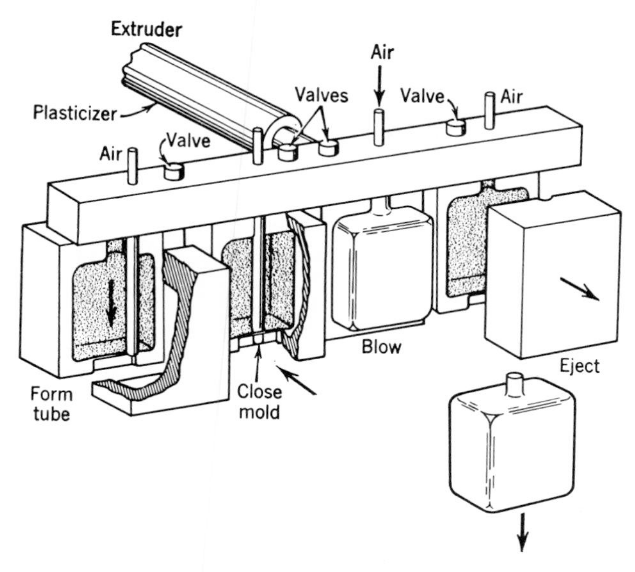

Fig. 10. Four-station manifold blow-molding system. Parison and mold action are controlled by timers so that each operates in sequence via valves in the manifolds.

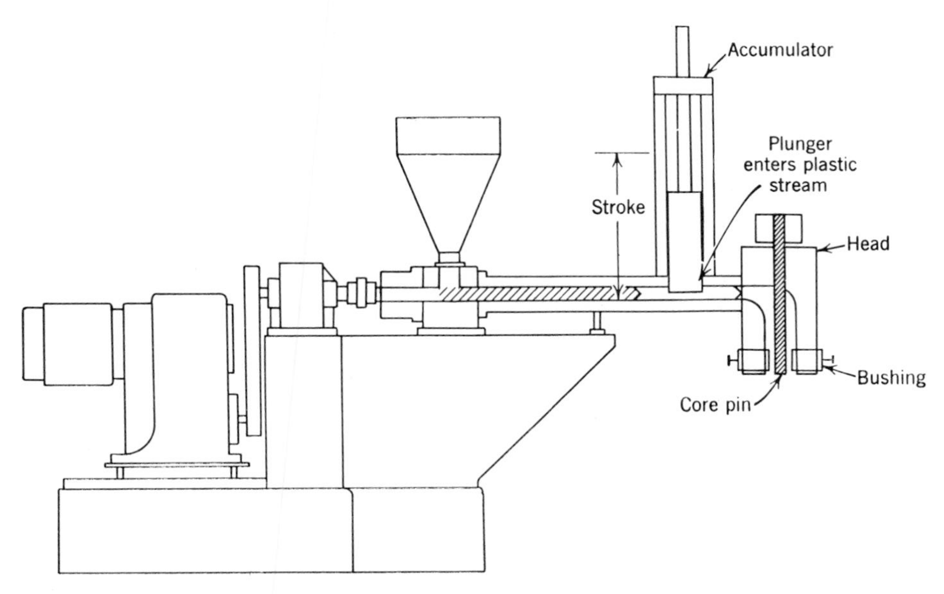

Fig. 11. Method of mounting an external ram accumulator at right angles to the barrel of the extruder. In this arrangement, the end of the accumulator enters the stream of plastic in front of the extruder end, cleaning the tip on each stroke.

rate stations, each with its own independent platen opening and closing mechanism. Figure 10 illustrates a four-station, continuous-extrusion valved manifold system. Parisons are extruded sequentially through an automatic valving arrangement in the manifold. Each parison is received by a fixed-position mold where blowing and cool-

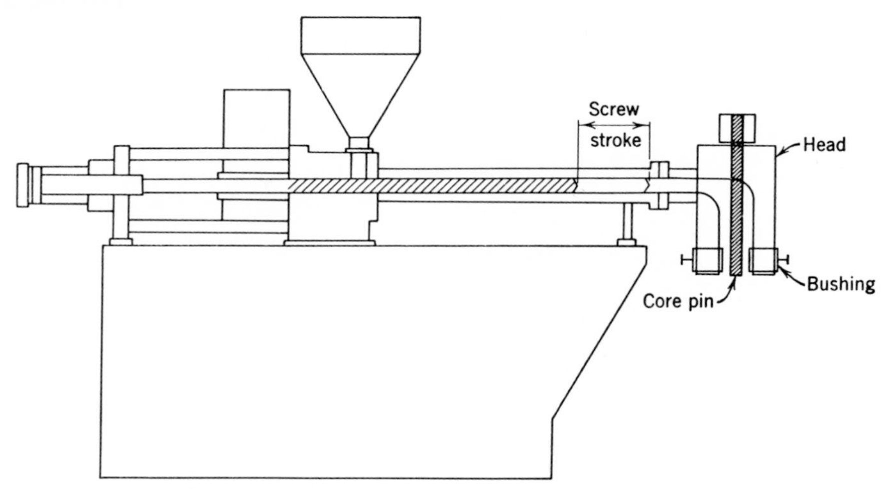

Fig. 12. Reciprocating-screw extruder. Functions as a combination of extruder and ram accumulator.

ing take place. An advantage of this type of operation is the ability to mold different items on a single machine. At each position a suitable parison pin and bushing would be used to match the desired product specifications. Separate timing arrangements are also required for each position. Long manifolds greatly increase back pressure on the melt and reduce cycle time. This type of machine is of a proprietary nature and is not available commercially.

With a two-station manifold system, a parison is extruded from one head, and when it reaches the proper length, the manifold valve automatically diverts the flow of material to the other manifold head. The mold closes on the parison and begins blowing and cooling the part while another parison is being extruded on the other side of the machine. After blowing and cooling, the mold opens, the part is ejected, and the mold is ready for a new parison.

Most machines of this nature have subsequently been converted to the accumulator extrusion type of operation using either external accumulators (Fig. 11) or reciprocating extruders (Fig. 12) to extrude parisons rapidly and reduce the total cycle time. The reciprocating-screw extruder has the advantages of straight-line flow of material as well as of using material on a first in–first out basis compared to the first in–last out basis of external accumulators.

Incorporating an accumulator or reciprocating screw into the blow-molding machine combines the principles of extrusion and of injection molding for the purpose of forming tubular parisons more rapidly. For accumulator operations, the extruder plasticizes material and feeds it to a separate cylinder or accumulator (Fig. 13). When a fixed level is reached in the accumulator, automatic valving stops the flow and the forward ram stroke forces material through the die orifice, forming the parison.

Reciprocating Molds. In Europe accumulators are not widely used; instead, most machine builders move the molds back and forth to the parison head in various manners permitting extrusion of parisons on a continuous basis.

In the case of vertical mold movement, the parison is extruded between the open mold halves. When the proper parison length is reached the mold closes, the parison is cut off by the mold, and the mold assembly moves downward to a fixed position beneath the extrusion die. During the downward movement the part is blown and cool-

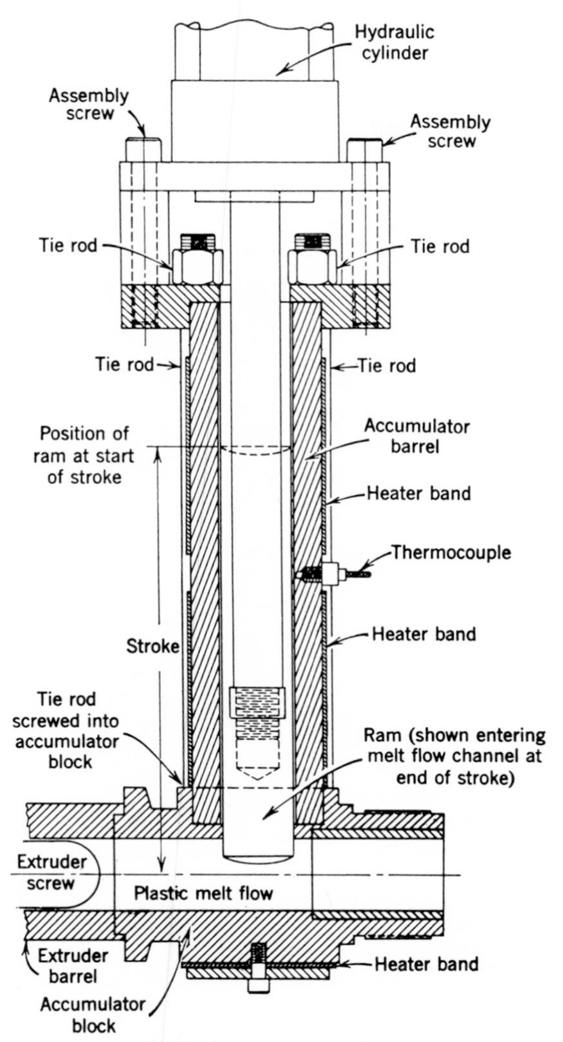

Fig. 13. Detailed construction of self-cleaning external ram accumulator mounted perpendicular to flow channel.

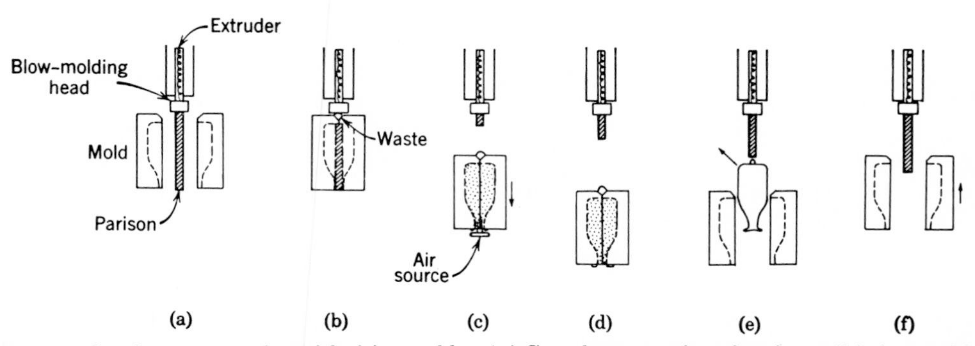

Fig. 14. Continuous extrusion with rising molds. (**a**) Complete extrusion of parison; (**b**) close mold; (**c**) blow; (**d**) cool; (**e**) eject; (**f**) raise mold.

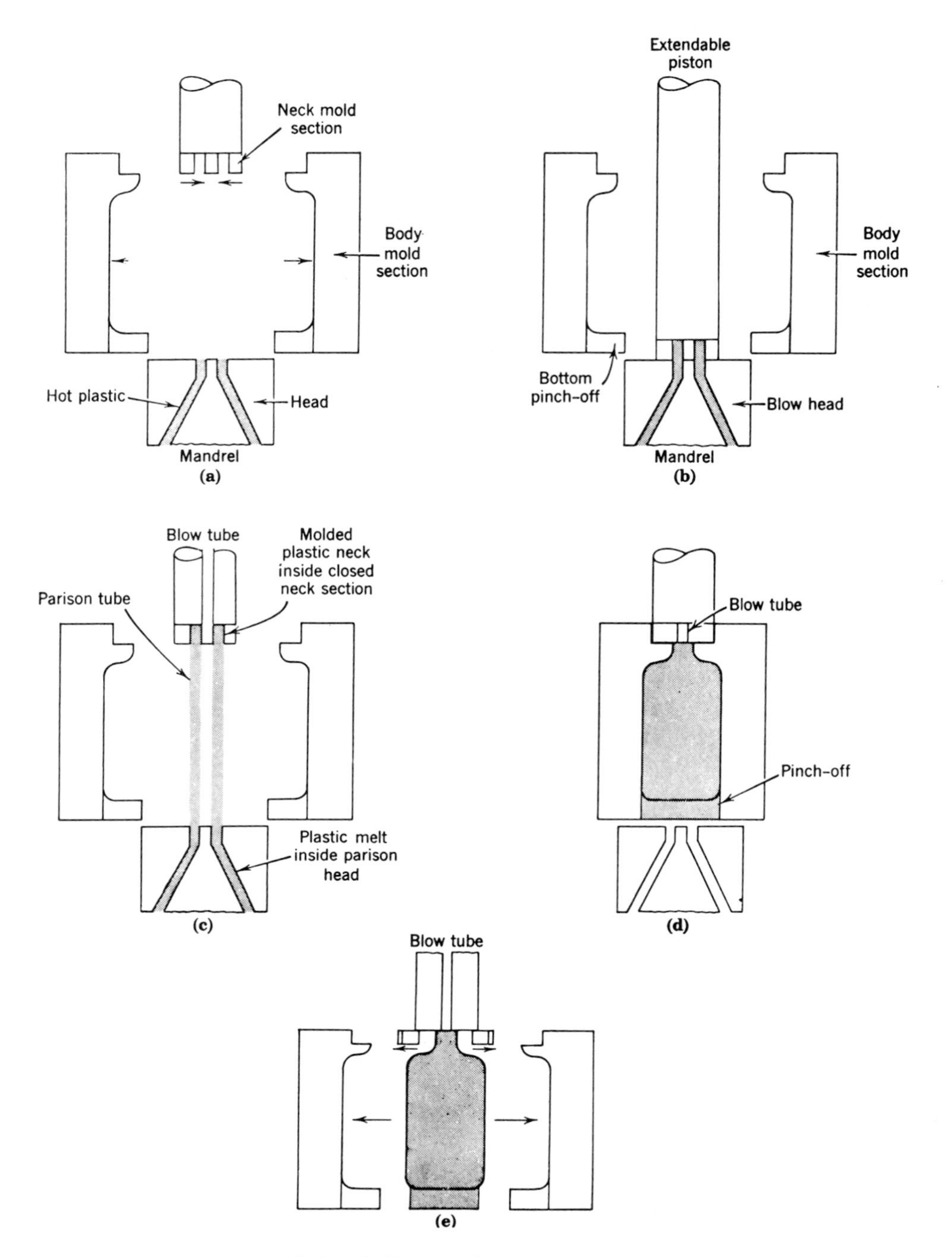

Fig. 15. Extrusion-molded neck blow-molding process. (**a**) Body section open, neck section closed, neck section retracted; (**b**) neck section extended to mate with parison nozzle (plastic fills neck section); (**c**) neck section retracted with parison tube attached; (**d**) body section closed, making pinch-off (parison blown to body sidewalls); (**e**) body molds open, neck molds open, bottle about to be ejected.

ing begins (Fig. 14). At the completion of the cooling cycle the mold opens, ejects the blown part, and automatically moves upward to repeat the cycle. Timing must be exact; slight fluctuations in timing produce parisons which are either too short or too long, resulting in a reject or in excessive waste in the tail pinch-off.

Most European vertical-movement blow-molding machines utilize a bottom blow pin over which the parison is dropped before the mold closes. Air is introduced through the bottom pin. By introducing cooling water to the bottom mandrel pin, cooling can be accomplished and cycles reduced. By careful sizing of the mandrel pin both the inside diameter and the outer surfaces of the bottle finish can be "calibrated" and a bottle can be produced which is substantially finished without external reaming operations (see p. 94). This requires removing the excess neck flash and the tail by mechanical action in the molds. This process is very popular in Europe and is now becoming popular in the United States. A pinch-off fin exists across the threads or on the sealing surface of the neck finish on "calibrated" bottles. Detailed tool design, however, has reduced this pinch-off fin to commercially acceptable levels.

Machines with horizontal mold movement are very similar to those with vertical mold movement. As the parison reaches the proper length, the mold halves close, the parison is severed at the die, and one mold or mold base containing multiple cavities moves horizontally to the right or the left. This method allows easy access to the top of the mold and is often combined with the use of an overhead driven blow pin to "calibrate" the neck opening; the driving mechanism is bulky and is difficult to mount above or below vertically moving molds.

Parison Transfer. For large items and multicavity blow molding, European machine manufacturers have developed a technique of transferring the hot parison vertically rather than by means of reciprocating molds. The approach that is used is to grab the parisons with mechanical fingers and move them vertically or horizontally to fixed mold positions. Such movement is subject to a certain amount of gravitational effect since the parisons continue to lengthen as long as they are held vertically. Vertical transfer of the parison creates slightly more parison drawdown than reciprocating mold processes. Horizontal parison transfer has not proved commercially acceptable owing to the sway of the parison during movement. The cost of vertical parison transfer mechanisms is considerably less than the costs of reciprocating platens or accumulator systems.

Extrusion-Molded Neck Process. An important proprietary blow-molding process used commercially is described in the patent literature as combining the features of a molded neck with conventional parison tube blow molding.

In this process (Fig. 15), a partible neck mold section is mounted separate from but capable of intermeshing with the main body of a blow mold. The blow mold is mounted in an upright position above an inverted parison head. The cycle starts with the main body of the mold open and the neck section in a closed position. The neck section is mounted so as to permit relative axial movement into and out of contact with the parison nozzle. When in position against the nozzle extrusion pressure fills the neck section with plastic melt. Axial return of the neck section at the proper speed synchronized with the rate of extrusion produces a parison attached to the molded neck and suspended between the open halves of the mold body. Closing of the body halves produces intermeshing of the neck section of the mold with the body section. Bottom pinch-off, blowing, and ejection then proceed as in conventional blow molding.

This process has the advantage of producing accurately molded necks with no chips or external parting lines. The extra step of molding the neck reduces the speed and increases the cost of the basic equipment. This is offset to a large extent by eliminating the requirements for finishing equipment and labor.

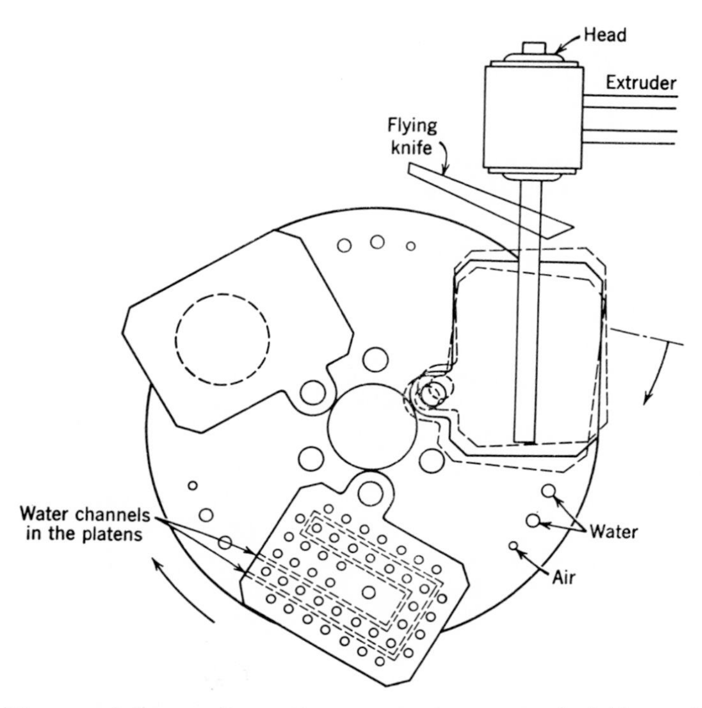

Fig. 16. Diagram of three-cavity continuous extrusion rotary-wheel blow-molding machine. Platens are shown without molds to illustrate water and air channels and platen position when clamping around the continuous parison. The molds can be timed to clamp exactly at vertical, or slightly before or after vertical, for blowing unsymmetrical items. Courtesy Beloit Eastern Corp.

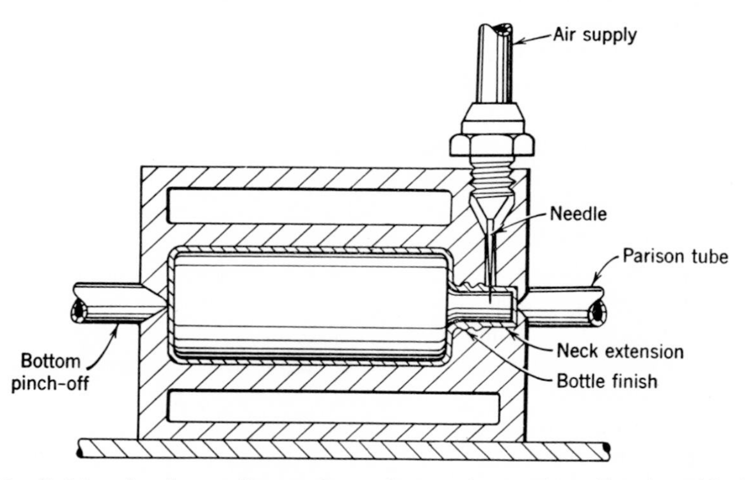

Fig. 17. Needle blow showing needle entering neck-extrusion portion of bottle mold. This portion is subsequently trimmed off.

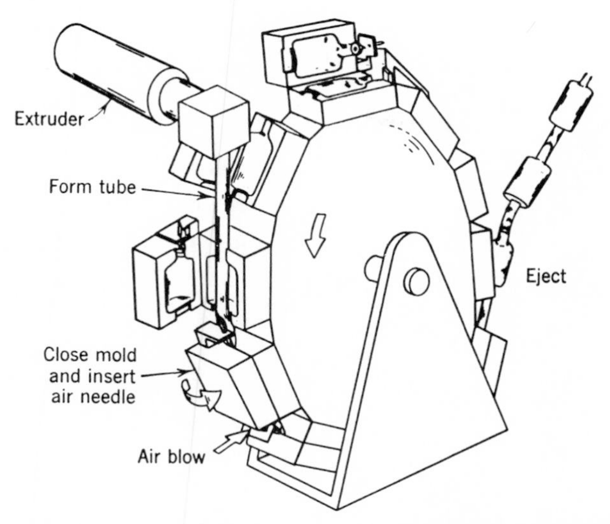

Fig. 18. Multicavity continuous-tube wheel-type blow-molding machine. Individual bottles are usually separated from the continuous parison tube at the time of mold opening.

Vertical Rotary-Wheel Blow Molding. This blow-molding method utilizes a continuous parison extrusion and a continuously moving vertical wheel, not unlike a ferris wheel. Rotation of the wheel can be either perpendicular or parallel to the axis of the extruder. Figure 16 illustrates a three-cavity rotary wheel. In this version the revolving speed of the molds exceeds the speed of the parison extrusion. When the molds clamp around the parison, the excess parison tube is severed by a flying knife. The molds then move away, leaving room for the extruding parison tube to form.

In the specific design shown in Figure 16, blowing is accomplished via an internal blow tube built into the mold. Increased air pressure is used to break through the hot plastic in the side wall of a neck extension. This neck extension is trimmed off during finishing operations. Needle blowing can also be carried out by piercing the parison in the neck extension area using either a fixed or driven pin, as shown in Figure 17.

Blowing can also be accomplished by the insertion of a calibrating mandrel (p. 96). When an external calibrating mandrel is used the mechanism is usually mounted on the top plate of the mold so as to slide parallel to the plane of the mold closing during the cycle in order to avoid physical contact with the blow head or the hot parison.

Figure 18 illustrates a multicavity vertical wheel. In this variation the speed of the tube and the mold movement are usually synchronized, resulting in a continuous string of bottles with pinch-off material still attached. In commercial practice the bottles are usually separated during the opening of the molds and the individual bottles are conveyed pneumatically to subsequent finishing equipment.

A two-cavity continuous-tube process has been developed in Europe utilizing a "hand-over-hand" mold-clamping mechanism (Fig. 19). The string of bottles is held together and moves in a festoon to the finishing station where the bottles are automatically separated and reamed.

One advantage of a continuous tube is that the bottom end of the tube can be closed early in the extrusion cycle and air can be introduced into the open end via the

Fig. 19. Hand-over-hand continuous-tube blow-molding machine equipped with automatic trimming, treating, and printing units. Completed printed bottles are deposited in final shipping container at the front center of the photograph. Courtesy Industrie-Werke Karlsruhe.

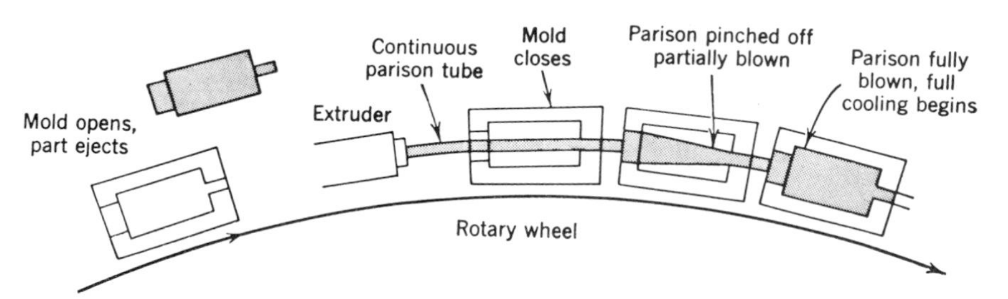

Fig. 20. Multicavity continuous-tube blow-molding machine.

mandrel pin. The tube is thus preblown early in the cycle. This technique is particularly useful in making offset handles on bottles and similar products requiring wide tube diameters almost equal to the distance across the parting line. When using free-falling parisons, wide tubes tend to wrinkle and crease with an adverse effect on wall-thickness distribution and the general quality of the product. Another advantage is the possibility of stretching the tube mechanically in the extrusion direction. This reduces the wall thickness of the parison and can sometimes produce a degree of directional orientation in certain plastics. Some biaxial orientation is possible if stretching in the machine direction is used in conjunction with preblow and temperature control. See also BIAXIAL ORIENTATION.

A disadvantage of the continuous tube process is the inability to use a calibrating mandrel for sizing and cooling. Also, in a machine in which the molds are at a fixed angle in the rotating wheel, the parison tube moves through a small arc as it moves from one mold to the next. Newer designs such as that shown in Figure 20 allow movement of the molds to position them close together in a straight line during extrusion and then back into a position slightly apart from each other for blowing. This movement permits straight-line extrusion, but the crowded condition on the wheel makes the use of a calibrating mandrel almost impossible.

Horizontal Rotary Machines. In this process (Fig. 21), plasticized material is conveyed to a single extrusion head. As the tube is extruded downward (on either a continuous or an intermittent basis) a rotary table mounted with molds is indexed below the extrusion head. When the tube reaches proper length, the mold closes, the tube is severed from the orifice by a cut-off knife, and the rotary table is moved to bring the next mold into position. Air is introduced into the tubing by a blowing mechanism, and the parison is held under pressure until the station ahead of the extrusion orifice. At this point, the blow mechanism is moved out of the way, the mold opens, and the bottle is discharged onto a conveyor. By using special molds the tails can be separated from the bottom of the container, leaving only top trimming to be accomplished by the finishing machine.

The horizontal rotary blow-molding machine uses a single-parison continuous extrusion, which is advantageous for processing heat-sensitive materials. The machine can be fitted with either or both top-blow, bottom-blow, and needle-blow mechanisms. Calibration pins can be used from either the top or bottom position. All actuating controls as well as air and chilling water are distributed to the individual molds via the center columns.

Because commercial horizontal rotary blow-molding machines operate in a "stop–start" manner, mechanical factors, particularly inertia, limit their speed to about 1 second per cycle.

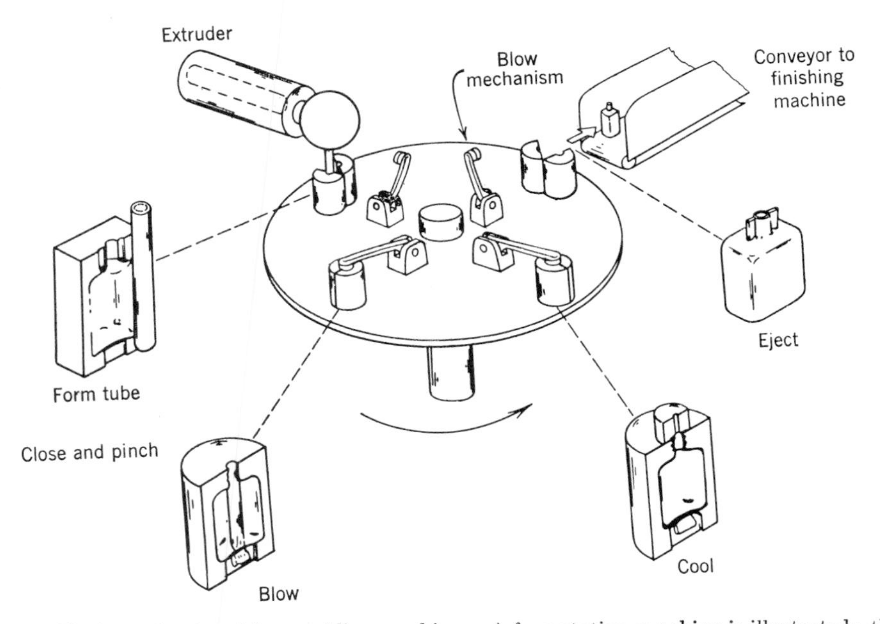

Fig. 21. Horizontal rotary blow-molding machine. A four-station machine is illustrated; the turn table indexes each mold intermittently.

Cold-Tube Process. In the cold-tube blow-molding process a cold parison, which has usually been produced at another location, is heated, positioned in a blow mold, and blown into shape. This process was designed specifically for poly(vinyl chloride), but is also being developed for use with other materials.

Reciprocating platens are used on which are mounted six, eight, or ten molds. The parisons are carried on a chain conveyor through a heating oven and then to the platen where the molds close upon them. After the parison is blown to shape and cooled, the molds open and the chain conveyor carries the finished containers to an ejection station. The empty conveyor section then moves on to the automatic reload position and the cycle continues. The ejected bottles are fed to a trimming machine or can be trimmed manually.

The cold-tube blow-molding process has the advantages that extruder production rates and conditions are independent of blow-molding rates and conditions; in addition, the open tube permits dissipation of gases generated during extrusion, whereas in continuous blow-molding processes these gases are held inside the blown container. On the other hand, the process does not lend itself to pinch-off handles, lugs, etc; moreover, reheating temperatures are lower than the parison temperatures in conventional blow molding, and as a result the bottom pinch-off is not as strong or as neat as on a conventionally blown bottle.

Injection Blow Molding

In the injection blow-molding process (Fig. 22) the parison is injection molded around a mandrel which is movable and contains a blowing tube (see also under Injection Molding. While it is still hot, the parison is transferred to a blow mold where air is introduced through the blowing tube in the mandrel. Items with more closely controlled dimensions can be produced by extrusion blow molding; however, their size is somewhat limited, principally by economic factors. The advantages and disadvantages of injection blow molding are summarized in Table 2.

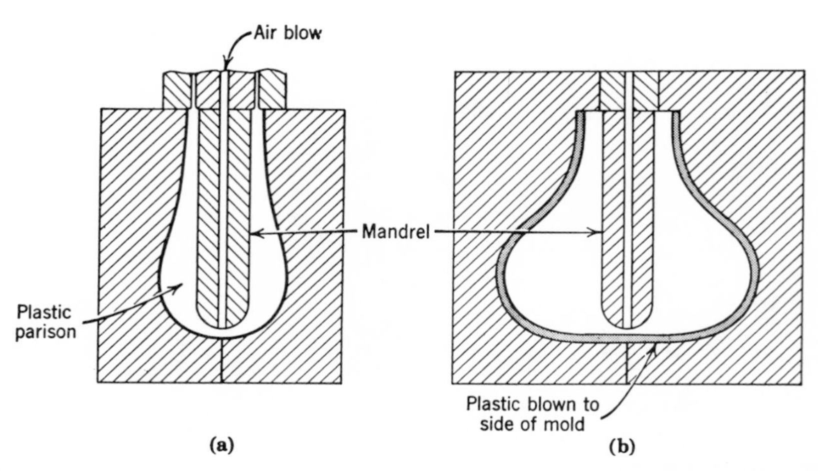

Fig. 22. Schematic drawing of basic injection blow-molding process. A plastic parison is injection molded onto a mandrel in a preform mold (**a**). The mandrel is transferred to a blow mold (**b**) where the parison is blown to final shape. Detailed shaping of the parison can be accomplished by contouring the mandrel or the preform mold or both.

Since the process requires two molds, as compared to one for extrusion blow molding, the tooling costs are considerably higher for conventional shapes, such as standard bottles; for special shapes that cannot be produced by either conventional injection molding or extrusion blow molding, tooling cost is not a deterrent. Machine costs for injection blow molding are also higher than for the less complex extrusion blow process. On the other hand, the only scrap produced by the injection blow-molding process is in the form of reject parts; there are no tops, tails, sprues, or runners requiring regrinding and reprocessing during each cycle. A threaded neck may be molded into the parison during the injection-molding phase of the process; closer tolerances and special shapes, such as undercuts, which are not available by extrusion blow molding, can be obtained.

Table 2. Advantages and Disadvantages of Injection Blow Molding

Advantages	Disadvantages
no finishing, trimming, or reaming of the blown item	double or triple molds per item
more uniform wall and neck thickness since all sections of the injection-molding parison are accurately molded	high cost of equipment
variable wall thickness of the parison to provide more resin for section to be blown to a greater diameter, making unsymmetrical items easy to blow	some injection strains in the blown item
no pinch-off or flash	two-step cycle usually requiring equal and simultaneous time for each step and resulting in some wasted cycle time
no regrind to upset plastic melt viscosity	
close tolerances and undercuts obtainable at threads in a bottle neck	
better clarity and gloss attainable with rigid materials such as polystyrene	
absence of chips in containers produced	
improved temperature control	

A less obvious but important advantage of the process is the absence of scrap material in the feed stock. The viscosity of plastic melt is thus not inclined to change owing to inconsistent utilization of regrind, as is commonly experienced with conventional blow-molding equipment, so that greater efficiency and more uniform products are obtainable. Formation of the initial parison between two metal parts makes possible close control of the temperature of the plastic to be blown, through proper temperature control of these metal surfaces. The uniform parison temperature results in uniform wall thicknesses in the blown products.

A properly engineered injection blow-molding machine lends itself readily to automated production as well as in-plant operations. The process can be readily standardized since it lacks the variations introduced by manual operations and scrap recovery. This process has been successfully used commercially in certain specialized fields with polyethylene and polystyrene. Many other materials, including poly(vinyl

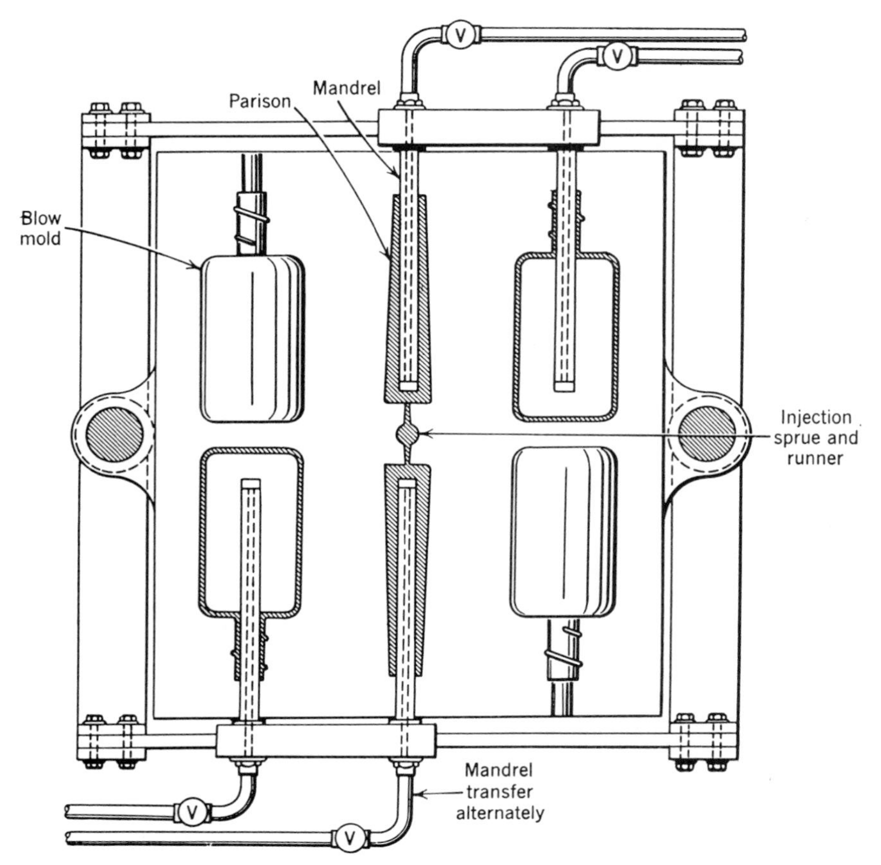

Fig. 23. Gussoni injection blow-molding process. Parison is injection molded on center mandrel while companion mandrel is in the blow station. Both mandrels are transferred simultaneously into juxtaposition with alternate blow cavities for final blowing. This process utilizes threaded neck sections in both injection mold and blow mold.

chloride), acrylonitrile–butadiene–styrene copolymers, and the acrylic resins, can be handled and have been molded to a limited extent by this process.

Equipment. A considerable variety of injection blow-molding machines has been described in the patent literature; some of the most significant of these will be described in this section.

A very complete, although somewhat complex, injection blow-molding machine was described by Canfield in 1954, in U.S. Patent 2,914,799. This machine is very similar in basic principle to one of the basic processes for producing glass containers. It consists of a multistation horizontal rotary injection-molding machine with the parison transferred to a similar rotary blow-molding machine. In 1957, in U.S. Patent 3,011,216, Gussoni disclosed another lateral transfer method consisting of core pins alternating between dual blow molds and a single injection mold wherein the blowing of one item is initiated before the injection molding of the next (Fig. 23). In 1960, in U.S. Patent 3,116,516, Moslo described a machine using four or more core pins transferring laterally between two sets of four or more blow molds and one set of four or more injection molds; blowing and injection occur concurrently in this machine.

A rotary machine of four stations described in U.S. Patent 3,100,913 filed in 1957 is shown in Figure 24. Injection takes place in station 1, simultaneous blowing in station 2, ejection in station 3, and core-pin cooling in station 4. In this process the

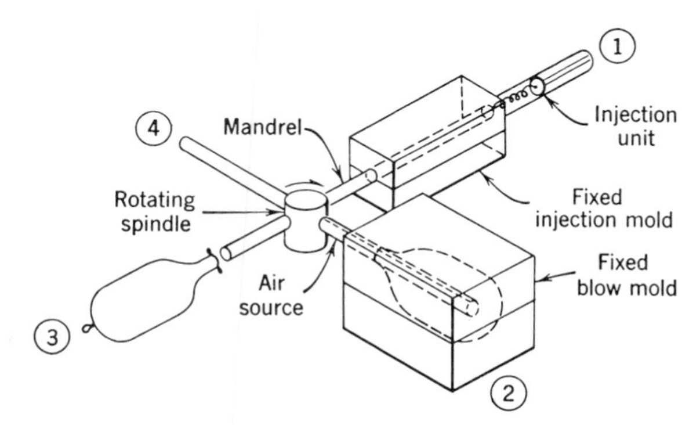

Fig. 24. Four-station rotary injection blow-molding machine. Key: 1, complete injection molding of parison; 2, close mold, blow, and cool; 3, eject; 4, cool mandrel.

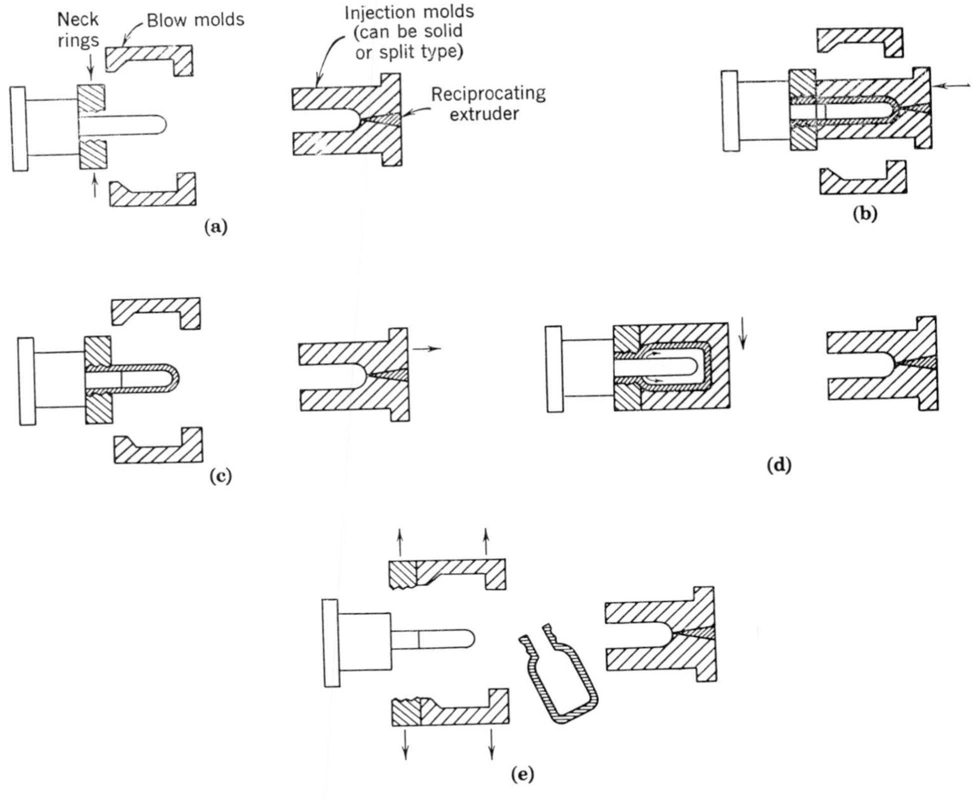

Fig. 25. Toshiba injection blow-molding process. (**a**) Neck rings close; injection molds close (if split type). (**b**) Injection mold advances over mandrel; plastic is injected into cavity. (**c**) Injection mold retracts (if split mold is used it must separate at this time). (**d**) Blow mold closes; blowing air enters parison via vents in mandrel. (**e**) Blow mold and neck rings open together; finished product is ejected.

single blow mold is in use during the entire time the single injection mold is being operated. A multistation rotary injection blow-molding machine was described by Saffron in British Patent 928,529 in 1960. In this machine the parison is injection molded above the first station onto a core pin extended from within the blow mold. The parison is lowered into the blow mold, and cooled and ejected at the other stations. This is an advance over the machines described previously in that it permits a very short hold time in injection followed by a long hold in the blow position. With simple modifications the Saffron process could also be used with a long hold in the injection mold for temperature conditioning of the plastic parison plus a single station for blow and cooling.

Toshiba Machine Company of Tokyo produces a two-station shuttling clump-type injection blow-molding machine, the operation of which is indicated in Figure 25. The machine uses a reciprocating-screw injection system which is advanced forward to mold the parison and then retracted as the molded parisons are enclosed by blowing molds for completion of the blow-molding cycle. Two sets of blow molds and mandrels are used in the machine. The clamp section containing the blow mold shuttles back and forth about the single central injection station to minimize loss of cycle time in waiting for a molded bottle to cool.

Blow-Molding Operation

The various problems most likely to be encountered in blow molding, along with their probable causes and the best approaches to a solution, are summarized in Table 3. Some of the most important factors in the process are discussed in detail in the following paragraphs.

The Extruder. In order to synchronize the rate of extrusion of the parison with the best possible blow-molding cycle, the extruder should have a drive with a continuously adjustable speed and should have an adjustment range in the ratio of at least 6:1. A hydraulic motor directly mounted to the extruder shaft will furnish maximum speed variation from zero to maximum rated speed. In case of overloading, a hydraulic motor will stall instead of causing damage in the event of a "freeze up" by the extruder. Mechanical motor drives should be positively connected to the main drive shaft of the extruder with a flexible coupling, an electromagnetic coupling, or a chain drive. These methods are generally preferable to a belt drive, which can slip and cause variation in speed.

Extruders with low length/diameter (L/D) ratios are used, but they produce a nonuniform parison and have a lower hourly capacity than longer machines. Longer machines permit a more gradual and uniform heat and energy input into the material. This produces parisons having more uniform viscosity, and also permits better control of stock temperature. Longer screws produce a minimum of surging and permit the use of lower cylinder temperatures to obtain the same melt temperatures, resulting in a very narrow range of stock temperatures. Narrow temperature distribution in the melt permits the production of precise and homogeneous parisons. Heat-sensitive polymers are extruded more readily on longer machines, since the cylinder temperature can be maintained at a lower level for a given stock temperature than with a shorter machine. Machines with high L/D ratios also provide better mixing with dry color blends (see also MELT EXTRUSION). An extruder for blow molding must deliver an extrudate with a relatively low stock temperature at high production rates. A breaker plate and screen pack are recommended to filter out any possible contaminants, in-

Table 3. Problems Encountered in Blow Molding

Problem	Probable causes	Correction
excess parison stretch	high melt temperature, slow extrusion rate	reduce stock temperature; high die temperature may be contributing factor; increase extrusion rate
rough parison surface	usually caused by low melt temperature; can also be caused by melt fracture or die fracture of material	reduce extrusion rate; raise parison bushing temperature gradually; if condition persists, check materials; try higher melt index resin; check die entrance angle; streamline overall die design
uneven wall thickness circumferentially in product	pin not centered in die bushing; uneven melt temperature; extrusion rate too high; faulty die design; unsymmetrical product shape	center die pin; check melt condition (probably too hot); reduce extrusion rate; check die design; shape die to increase thickness; use preblow; use larger parison
uneven wall thickness lengthwise in product	parison necking down; irregular product shape	reduce parison necking by increasing extrusion rate and lowering melt temperature; extreme case requires programmed control of parison extrusion
bubbles in parison	moisture causes many small bubbles; trapped air causes larger bubbles	for moisture, dry pellets; for air, increase screw speed; increase screen pack to increase back pressure
streaks in parison	contamination from "hang-up" in equipment; melt overheated and degrading material	clean die head; if condition repeats with clean system, check flow channels for hang-up areas; check for hot spots
parison forms into a roll	pin too hot, bushing too cold; parison clings to cooler surface	increase bushing temperature; cool pin
poor gloss on product	poor melt flow	try resin with higher melt index
ripples in parison, grooves or lines in product	contamination in die orifice; possibly poor die design; extrusion rate too high	clean orifice; check die design: check for nicks; reduce extrusion rate
product surface shows pits, fish scales, etc; parison smooth	trapped air in mold; condensed water on mold surface; insufficient blow pressure	vent mold, sand blast surface for polyethylene; polish surface for poly(vinyl chloride); increase size of air line and blowing pressure; raise die temperature to reduce condensation; check for water leaks; if condition is localized, check for air leak around the blow pin

Table 3 *(continued)*

Problem	Probable causes	Correction
product warpage	improper cooling	check uniformity of die cooling; wall thickness variation of product may be excessive, may require redesign or excessive cooling cycle; stress crack potential increases under conditions producing warpage
container breaks on weld seam	inadequate weld; melt temperature too low	raise melt temperature
thin wall streak at parting line	dies not completely closed; material stays hot at parting line; blow pressure causes stretching	increase clamp pressure; check to be sure dies do not bounce on contact
black specks	material contaminated or hold-up material flaking off melt channel	check incoming material; clean melt channel
part sticks in mold	mold temperature too high, cycle too short	improve mold cooling, increase cycle
parison pinch-off sticks to product	parison tail too long	shorten parison; provide additional pinch-off; relieve area to cool tail
parison blow-out	blow-up too rapid; melt temperature too high; thin section in parison due to contamination; pinch-off could be too sharp or too hot; product may have too high a blow-up ratio	program air blow-up start with low pressure and increase; check parison condition and temperature; check pinch-off; if blow-up ratio is too high, use larger parisons
excess shrinkage	poor cooling; melt temperature too high	increase blow pressure to obtain better cooling contact with mold surfaces; improve cooling in mold; increase cycle; reduce melt temperature
thin wall at pinch-off	pinch-off too sharp; inadequate blow; poor venting	increase pinch-off land width; reduce relief area to hold and cool pinched off material
thick wall at pinch-off	pinch-off clearance too high; pinch-off angle too small	reduce pinch-off clearance; open pinch angle to about 30°
undercuts fail to strip	product overcooled; shrinkage excessive; undercuts too severe	reduce cycle, relieve undercut; use movable die insert to product undercut

cluding large unplasticized particles referred to as gels. A screen pack containing at least one screen as fine as 100 mesh is recommended to ensure elimination of these foreign particles in the final blown product and to help homogenize the plastic melt. An extruder suitable for blow molding should be equipped with a thrust bearing suitable for 4000 to 5000 psi working pressures. Working pressures should be monitored by a pressure gage. The extruder should be equipped with a hard-surfaced screw, and the cylinder should be heavy walled and lined. The extrusion head and gate assembly must also be designed for high-pressure operation.

Back pressure improves extruder performance and mixing efficiency of the screw. This results in a more uniform and homogeneous parison. Increased extrusion pressures can be obtained by fixed orifices or restrictions, by screen packs, or by adjustable valves. Screen packs develop a relatively low, nonadjustable pressure which varies during production in an uncontrollable fashion owing to clogging. A fixed orifice or restriction can sometimes be satisfactory, but it lacks versatility, and an adjustable valve or orifice between the head of the extruder and the die is advisable.

Melt Temperature. The melt temperature in extrusion blow molding is very important and must be carefully controlled. Stock temperature is measured and indicated by means of a thermocouple immersed in the melt stream. High stock temperatures produce blow-molded articles with better finish as well as better physical and mechanical properties. Parisons at high temperatures, however, neck down because of gravity pull. High stock temperatures also require a longer cooling cycle and cause more shrinkage. Lower stock temperatures result in better shaped parisons, shorter cooling cycles, and less shrinkage. Too low a stock temperature produces a nonhomogeneous melt as well as inferior mechanical and physical properties in the finished product; it can also cause poor seals along the pinch-off areas. For best product quality, it is desirable that the parison have the lowest practical stock temperature.

The Parison. A uniform parison in blow molding is of paramount importance. The extruder must deliver an extrudate within close dimensional tolerances and without weight or viscosity variations. Irregular extruder output causes weight variations from one blown part to the next. Too heavy a part uses extra material; too light a part is mechanically faulty. A variation within the parison may produce a part with correct weight, but one that is too thin at a critical point such as at the shoulder or the bottom of a bottle.

The temperature of the parison die should usually be controlled so as neither to add nor to remove heat from the mass of the plastic melt. Thermocouples should be lo-

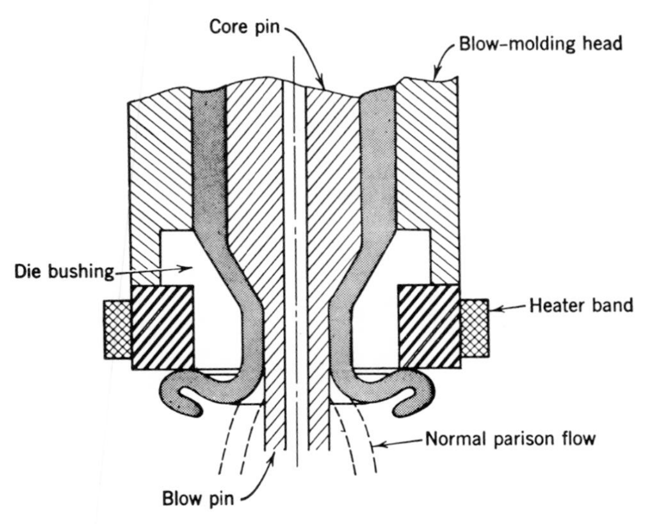

Fig. 26. Diagram showing parison curl, which occurs when bushing is too cold or core pin is too hot. Plastic moves away from the hotter area toward the cooler. If the temperature conditions were reversed, the plastic would tend to cling to the blow pin.

cated in the body of the die to permit fast sensing of temperature changes. The exit end of the parison die, referred to as the nozzle, should have a separate narrow-band heater. Operating with the nozzle heater hotter than the parison die will add luster to the surface of the parison. Care must be taken to heat only the surface of the parison and not change the temperature of the bulk of the plastic. On the other hand, if the nozzle temperature is too low the parison will usually curl outward around the nozzle instead of forming a proper tube. Effects of poor temperature control are shown in Figure 26.

Parison "drawdown" results from gravitational pull on the plastic during extrusion. The bottom becomes heavier while the top thins down. The lower the melt viscosity the greater the drawdown; drawdown can be reduced by lowering the melt temperature. Fast extrusion reduces drawdown because the material is acted on by gravity for a shorter time. Parison programming can be used to compensate for drawdown.

Parison die swell occurs upon exit from the parison die, and is caused by the release of compression created in the hot melt while passing through the parison die orifice. Within narrow limits parison swell can be used to increase or decrease parison wall thickness and thus alter container weight. Increased parison delivery speed increases swell, reduces drawdown, and results in an increased wall thickness in the finished product. Increased melt temperature has the opposite effect. Table 4 lists average percentage swell experienced with some common blow-molding materials.

Blowing Air. The air used for blowing serves to expand the parison tube against the mold walls, forcing the material to assume the shape of the mold and forcing into the surface details such as raised lettering, surface designs, etc.

During the expansion phase of the blowing process it is desirable to use as high a volume of air as is available so that the expansion of the parison against the mold walls is accomplished in the minimum amount of time. The maximum volumetric flow rate into the cavity at a low linear velocity can be achieved by making the air inlet orifice as large as possible. In the case of blowing inside the neck this is sometimes difficult. Small air orifices may create a venturi effect, producing a partial vacuum in the tube and causing it to collapse. If the linear velocity of the incoming blow air is too high, the force of this air can actually draw the parison away from the extrusion head end of the mold. This results in an unblown parison. Air velocity must be carefully regulated by control valves placed as close as possible to the outlet of the blow tube.

The pressure of the blowing air will vary the surface detail in the molded item. Some polyethylene items with heavy walls can be blown and pressed against the mold walls by air pressures as low as 30–40 psi. Low pressure can be used since items with

Table 4. Swell for Various Plastics[a]

	Swell, percent
polyethylene, linear	
Phillips type	15–40
Ziegler type	25–65
polyethylene, high-pressure, low-density	30–65
poly(vinyl chloride), rigid	30–35
polystyrene, general-purpose	10–20
polycarbonate	5–10

[a] Amount of swell varies widely depending upon actual operating conditions.

heavy walls cool slowly, giving the polymer more time at a lower viscosity to flow into the indentations of the mold surface. Thin-walled objects cool rapidly; therefore the plastic reaching the mold surface will have a high melt viscosity, and higher pressures (in the range of 50–100 psi) will be required. Large items such as 1-gal bottles require increased air pressure (100–125 psi). The plastic has to expand further and takes longer to get to the mold surface. During this time the melt temperature will drop somewhat, producing a more viscous plastic mass which in itself requires more air pressure to reproduce the details of the mold.

The clamping force of the mold platens must be sufficient to withstand the blowing pressure inside the blow molds. An excess of 25–50% pressure over the calculated value of projected area of the product times blowing pressure is advisable.

The use of atomized water or liquid carbon dioxide mixed with the blowing air for blowing parisons has resulted in reduction of molding cycles by up to 50% through the rapid cooling produced. However, the sudden shock of these methods of cooling can create strains in the plastic and can spot or pock mark the inside surface of the blown parisons.

Die Lines. Die lines are surface imperfections or scratches that are formed on the parison as it is extruded, and are transferred to the blow-molded article. The cleanliness of the parison die is important in reducing these lines. Chrome plating of the die or coating with Teflon can be helpful. Some resins, particularly high-density polyethylene, will produce die lines even in a clean die. Die lines can be the result of nonuniform melt viscosity. This can be reduced by high-density mixing and use of polymers of uniform molecular weight.

The effect of die lines can be minimized by any action that tends to minimize the cooling of the melt until it is pressed against the mold surface. These include (a) increasing blowing speed so that localized areas contact the mold before the rest of the piece is filled out; (b) increasing air pressure to fill the mold more rapidly and to flatten out the die lines; and (c) increasing mold temperature to reduce chilling the parison so that it is still pliable when high pressure develops in the mold.

Shrinkage. Lengthwise shrinkage in blow-molded products tends to be slightly greater than transverse shrinkage. Most of the horizontal shrinkage occurs in wall thickness rather than in body dimension. In the case of polyethylene, higher shinkage occurs with the higher-density polymers and thicker-walled products. The former is due to greater crystallinity of the more linear polymer, and the latter to the slower cooling rates which result in more orderly crystalline growth.

Shrinkage may be reduced by raising the blowing pressure and lowering the temperature of the mold. Rapid cooling is desired, but cooling too rapidly can cause surface imperfections and distortion. Raising the stock temperature, while it may not appreciably affect outside dimensions, causes more of the shrinkage to occur in wall thickness, because the higher melt temperatures lessen strain recovery and reduce blowing stresses.

Product Design. Blow-molded products should be designed with generous radii at corners and edges. Fillets and rounds should be employed wherever possible in corners, ribs, and edges. Such parts will possess more uniform wall thickness, and as a result of more uniform and faster cooling, internal stresses, as well as product distortion, will be reduced.

A generous radius in the chime area of bottles is important. Thinning causes stress cracking between the heavy pinch-off sections and the thinner edges. Rounded bottom edges contain lower levels of stress than do sharp edges and, therefore, have more resistance to stress-cracking agents. A concave base in bottles acts as a shock absorber and improves impact strength. The entire bottle design should be as elastic as possible so that impact energy can be absorbed. For this reason cylindrical bottles are preferred to containers of rectangular cross section, and the joint between the side wall and the base should have as large a radius as possible. The same principle applies at the shoulder side-wall and the shoulder neck joint. In waisted bottles a gently curved waist is better than a sharp, well defined waist. Sharp corners must be avoided at the attachment of any vertical or horizontal ribbing.

Large undercuts should also be avoided. These may prevent removal of the product from the mold or cause excessive distortion during removal. When large undercuts are necessary, it may be possible to provide for their removal with moving parts in the mold design (see MOLDS). The ratio of the diameter of the finished part to that of the parison should be in the range of 4:1, the exact value depending upon material, process, and design details.

Varying degrees of wall collapse may occur. This is due to a negative pressure within the container and may arise from diffusion of the contents outward through the wall, from absorption of one or more of the atmospheric gases within the bottle, or from premature sealing of a container which has been filled at an elevated temperature. By proper design, a certain amount of wall collapse in containers can be accommodated without noticeable distortion.

Auxiliary Equipment. To operate a blow-molding line the following additional equipment will be required: air compressor; hopper loader; scrap grinder; two mold-temperature controllers, one heater, and one chiller; trimmer and reamer; and dry blender. If, in addition, finished and/or decorated items are to be produced, surface treating equipment, labeling and printing equipment, and conveyors will be required. A detailed discussion of auxiliary equipment for molding is given in the section of this article on Auxiliary Procedures.

Testing Plastic Containers

Since by far the most common products of blow molding are containers of one sort or another, the testing of these articles will be described briefly here (see also CONTAINERS). In addition to checks for deviations from the required dimensions and for the defects of appearance of most plastic products, there are four tests of significance to blow-molded products, and particularly to containers. For blow-molded items other than containers, the tests employed will be determined by the intended use of the part.

Permeability. Permeability is ordinarily determined either at room temperature or, in accelerated tests, at 105°F. The weight loss in grams in a given time interval is determined and the extrapolated weight loss per year then calculated. The arbitrary value of 3% loss per year is considered as maximum for an acceptable package.

Environmental Stress-Cracking Resistance. Some chemicals, such as many surface-active agents, can cause environmental stress cracking. A simple accelerated test to determine stress-cracking resistance involves filling the container with the material in question and subsequently aging in a circulating air oven at either 125 or 150°F.

Impact Strength. To determine whether the container has sufficient impact resistance, a simple bottle-drop test can be employed. The container in question is filled with liquid, usually water, and allowed to drop from predetermined heights. The testing of impact strength is discussed in detail in the article IMPACT RESISTANCE.

Bibliography

"Blow Molding—Part 1," *Brit. Plastics* **32** (2), 48–57 (Feb. 1959).

"Blow Molding—Part 2," *Brit. Plastics* **32** (3), 97–103 (March 1959).

"Blow Molding—Progress Report," *Preprint, Soc. Plastics Engrs., Regional Tech. Conf., Hartford, Conn., Nov. 7, 1963.*

"Blow Molding Comes of Age," *Preprint, Soc. Plastics Engrs., Regional Tech. Conf., Newark, N.J., Nov. 18, 1960.*

"Blow Molding Equipment," *Plastics World* **17** (6) (June 1959).

"Blown Twin-Walls Bring Case Versatility," *Mod. Plastics* **42** (10), 107 (June 1965).

"European Processors Pave Way in Blow Molding Rigid PVC," *Plastics Technol.* **11** (1) (Jan. 1965).

"New Bounce for Plastic Bottles," *Mod. Packaging* **40** (6) (June 1966).

"PVC in Injection and Blow Molding," *Preprint, Soc. Plastics Engrs., Regional Tech. Conf., New York, Nov. 18–19, 1965.*

G. S. Brown, "The Mechanical Processes of Blow Molding," *Plastics World* **17** (5) (May 1959).

G. H. Burke and G. C. Portingell, "Melt Extrusion Properties of Rigid PVC," *Brit. Plastics* **35** (5), 254–258 (May 1963).

R. Doyle, D. E. Perry, and T. E. Branscam, *Extrusion Blow Molding Techniques for High-Density Polyolefins*, Phillips Chemical Co.

M. Ferar, "Blow Molding," *Ind. Design.* **7** (9) (Sept. 1960).

R. E. Hartung, "Molds for Blow Molding," *Mod. Plastics* **34** (7), 157–222 (March 1967).

V. Hill, "Blow Molding—Comparison of Principles Related to Economics and Markets," *Mod. Plastics* **41** (11), 89–92 (July 1959).

F. W. John, "Blow Molding Fundamentals," *Plastics World* **24** (7), 54–58 (July 1966).

D. Jones and T. Mullen, *Blow Molding*, Reinhold Publishing Corp., New York, 1962.

S. Ohtsuka, S. Yoshikawa, and Y. Hoshi, "Internally Plasticized Ether Copolymer Resins for Injection and Blow Molding," *SPE J.* **22** (5) (May 1966).

E. Overgage and D. Burgess, "Successful Blow Molding of High-Density Polyethylenes," *Brit. Plastics* **36** (6), 328–336 (June 1963).

J. H. Parliman, "Blow Molding Plastic Bottles," *Packaging Eng.* **00,** 62–69 (March 1967).

R. Platte, " Materials Guide: Designing Blow-Molded Containers," *Ind. Design* **14** (4), 58–62 (May 1967).

S. Rupert and B. Strong, "Blow Molding," *Mod. Plastics* **42** (7), 99–101 (March 1965).

R. L. Wechsler and T. H. Bagles, "Blow Molding Polyethylenes—Parts 1 and 2," *Paper, Soc. Plastics Engrs., ANTEC Meet., New York, Jan. 27–30, 1959.*

H. A. Williams, "Production and Properties of Rigid PVC Bottles," *Brit. Plastics* **37** (4), 198–201 (April 1964).

G. E. Pickering
Arthur D. Little, Inc.

ROTATIONAL MOLDING

Rotational molding (also referred to as rotational casting) is, as a processing system, not a new technique, having had a place in industry for approximately twenty years. Initially, and over a period of the first fifteen years, rotationally cast items were fabricated almost exclusively with vinyl plastisols. The applications were somewhat limited, however, and consisted mainly of novelties and decorative items, such as artificial fruits, mannequins, and other hollow display figures, and children's toys. In 1961, the first polyolefin powder, polyethylene, was publicly demonstrated to the rotational-molding industry. It was quickly realized in the industry that polyolefin powders offered new applications of rotational molding in the plastics field, many of which would be virtually impossible by any other method. Rotational molding of polyolefin powders is distinctively different from other plastics molding processes, such as blow molding and injection molding, in that it: (a) uses resin powder instead of resin pellets; (b) melts the powder inside the mold instead of forcing molten resin, under pressure, into the mold; (c) uses biaxial rotation of the mold; and (d) utilizes molds that are comparatively inexpensive because of their simplicity, due to the lack of pressure in processing and lack of coring for water cooling. Rotational-molding techniques involving polyolefin powders have been developed so that in the production of many items they can compete with blow and injection molding and with thermoforming. See also other sections under Molding: Blow Molding; Compression and Transfer Molding; Injection Molding.

Advantages

Rotational molding created not only a technique to broaden the limits of the design of an article, but also has many advantages when compared with other molding techniques: (a) Costs for molds and tooling are extremely low, especially for short runs. However, the technique is easily adapted to continuous production methods and high production rates, particularly when multiple-cavity molds are used. (b) Prototype molds can be manufactured for experimental purposes without great expense. (c) In most cases, rotational molding permits elimination of secondary tooling. Frequently, producing items in one piece eliminates secondary operations. (d) There is little or no waste scrap; the desired weight of powder is placed into the mold. This, in turn, closely controls wall thickness and piece weight of the molded part. (e) Pieces with undercuts and intricate contours can be easily molded, limited only by practical size limits of the mold and oven and the strength of the mold spindles. (f)

Cross-sectional deformation and warpage are at a minimum. (g) Relatively stress-free items are produced owing to a minimum of "frozen-in" stresses. This protects articles from stress cracking when exposed to environments inducing cracking (see Long-Term Phenomena under FRACTURE). (h) Identical or similar items or different sections of one piece can be molded at the same time in different colors on a single spindle. (i) Double-wall constructions are feasible (Fig. 16). In addition, pieces with excellent detail and finish can be produced and any type of grain can be obtained on the exterior surfaces. Dimensional tolerances of the side walls and bottom of the moldings are easily controllable. Since rotational molding, by the nature of the procedure, assures uniform wall thickness, deviations can be controlled to within very close tolerances. Plastic or metal inserts can often be molded as integral parts of the item. The lack of weld lines adjacent to inserts or openings improves impact strength of the finished item.

Resins

The introduction of powdered polyethylene for rotational molding provided fabricators with a choice of materials for their operation since essentially the same equipment can be used for processing either vinyl dispersions or powdered polyolefins. When both of these types of resin are suitable for a product, the fabricator must study raw material costs, production rates, and end-product properties before making the resin choice.

Vinyl Dispersions (qv). The rotational casting of liquid vinyl dispersions, or plastisols, has a sizeable production volume consisting mainly of toys and novelties, such as dolls, beach balls, artificial fruit and flowers, as well as functional items such as toilet balls, squeeze bulbs, and housings for battery testing units. More recently, automotive armrests and safety sun visors have been rotationally cast with vinyl plastisols and filled with polyurethan foam.

The selection of the plastisol for a specific application is determined by the physical and/or chemical properties required by the end product. A suitable plasticizing monomer is then incorporated into the poly(vinyl chloride) to formulate the casting liquid. The liquid-to-solid conversion in rotational casting takes place around 350–400°F (175–205°C) for most vinyl plastisols.

Powdered Polyolefins. The use of a high-quality polyethylene powder is essential in rotational molding if good end products are to be obtained. However, the high temperatures used produce the risk of chemical degradation. Therefore, powdered polyethylenes are offered with a variety of combinations and wide ranges of melt indexes, densities, and particle sizes. The effects of melt index and density on the end-product properties and processability are quite critical for rotational molding, whereas those of particle size are considerably less (Table 1). See also ETHYLENE POLYMERS.

Melt Index. Polyethylenes with fairly high melt indexes are generally more suited for applications where high production rates are a primary requisite, such as rotational molding of low- and medium-density powders, than are polymers with low melt indexes. However, maximum toughness, ie, good impact strength as well as resistance to low-temperature brittleness, is obtained with powders with lower melt index. These powders are thus more suited to rotational molding of linear polyethylene powders; a lower melt index also increases toughness of high-density resins.

Table 1. Effect of Increases in Melt Index and Density of Polyethylene Powder on Processing and Properties

	Melt index increase	Density increase
melting point	decreases	increases
flow	increases	
impact strength	decreases	decreases
stiffness		increases
Vicat softening temperature	decreases	increases
resistance to low-temperature brittleness	decreases	decreases
barrier properties		increase

Density. Generally speaking, low-density resins are preferable when stiffness is not essential or when it is undesirable, as for many toys, and where only light loads are expected. Medium-density resins are useful for self-supporting items which require higher heat-distortion resistance or stiffness which low-density resins will not provide. High-density resins impart highest rigidity to the end product, which frequently permits reduction of the wall thickness and therewith of the cost per piece. In addition to increasing stiffness, increasing the density raises the melting point, permits higher service temperature limits, and improves barrier properties of the end product.

Basic Molding Steps

There are six basic steps associated with the powder molding of a finished product:

1. The cavity, or multiple cavities, in the bottom parts of the unheated molds (Fig. 1) are charged with a predetermined weight of powder desired in the end product.

2. The mold parts are clamped in place. For totally enclosed pieces, the entire mold is made of heat-conductive material. When one or both ends of the pieces are open, heat-insulating covers are used to close the mold.

3. The charged molds are placed in an oven, where they are heated (Fig. 2) while simultaneously rotating around two axes in planes at right angles to each other. Figure 3 shows the principle of such biaxial rotation.

4. The double revolving motion results in formation of hollow objects in every mold cavity, the powder being evenly distributed to form walls of uniform thickness when the resin fuses, except where heat-insulating covers are used. Weight and wall thickness of the molded items can be modified by increasing the amount of powder initially put into each lower mold cavity.

5. When all the powder has fused into a homogeneous layer on the walls of the cavities, the mold is cooled while still being rotated (Fig. 4).

6. The mold is opened (Fig. 5) and the molding is removed. Then the mold is readied for the next cycle.

Molds

A mold required for rotational molding can be simple because very little pressure is exerted in the process and no coring for water cooling is necessary. Therefore, such a mold costs only a fraction of what a comparable injection or blow mold would cost.

Fig. 1. Left: Opened fifteen-cavity mold used for rotational molding of balls. The bottom cavity halves are ready to be charged with powdered polyethylene. The mold is hinged for easy opening, filling with powder, and, after processing, removal of the molded products. Right: A multi-cavity mold like the one at left is just entering the oven at the beginning of the heating cycle.

Table 2. Heat Conductivities of Metals Used for Making Rotational Molds

	Btu/hr/sq ft/°F/in.	g-cal/hr/sq cm/°C/cm
steel	26	31
aluminum	124	148
copper	215	257

Split molds are required for piece removal when rotationally molding powders. If the very flexible vinyl resins are being used, the part may be removed from a one-piece mold through a relatively small hole. The selection of rotational molds depends mainly on the size, shape, and surface finish of the piece to be molded and also on whether one or several molds are to be made for a particular piece. Molds should be as thin-walled and as light in weight as practical. See also MOLDS.

Mold Materials. Possible materials for construction of molds are sheet steel, cast aluminum, or electroformed copper–nickel alloy. Aluminum or electroformed copper–nickel alloys are recommended for small or intermediate-sized molds, complex

Fig. 2. The charged mold is heated in an oven, where it rotates biaxially.

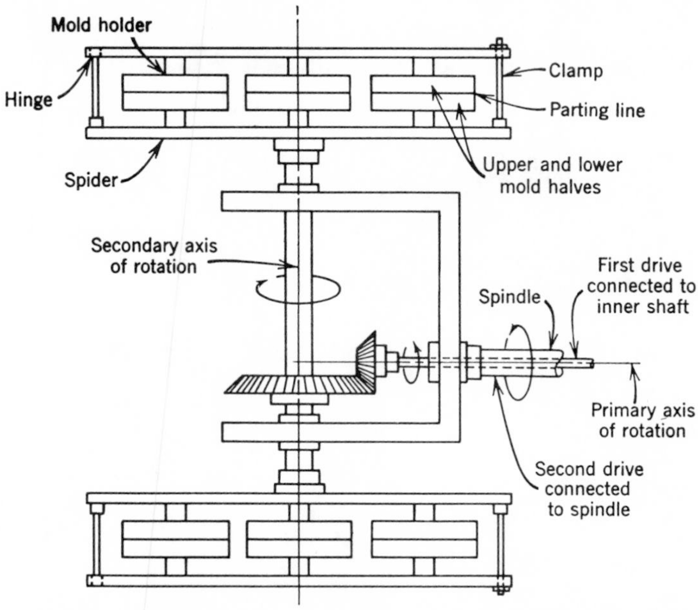

Fig. 3. Mechanism of biaxial rotation. The spindle is turned on the primary axis while the molds are rotated on the secondary axis. Usually, the ratio of the two simultaneous rotations is determined by the gear, which may be exchangeable, or by two motor drives.

shapes, and when several molds of one design are required. Steel should be used for most large molds and for a prototype mold of a single design because it is the most economical material. Table 2 lists the heat conductivities of metals used for making rotational molds.

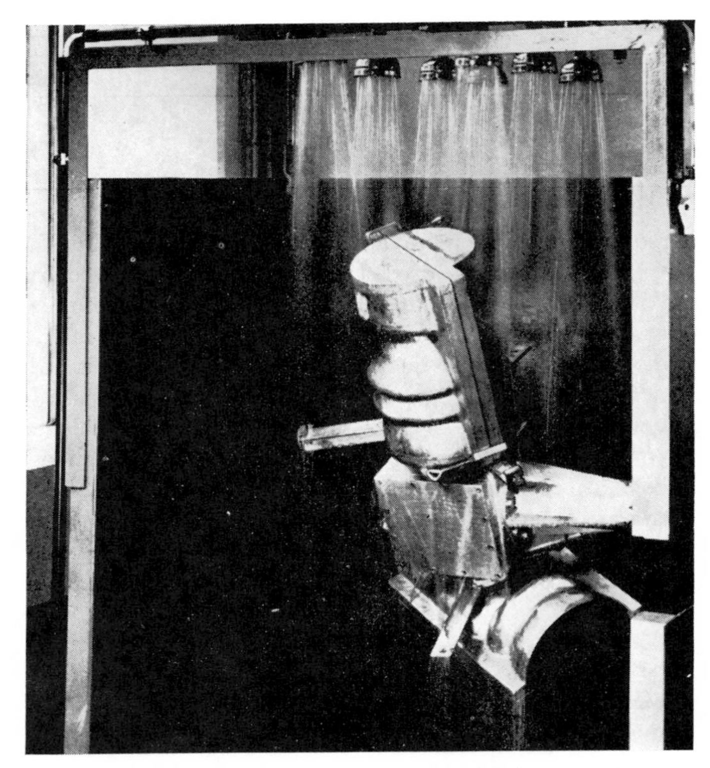

Fig. 4. Cooling the mold while it is still being rotated.

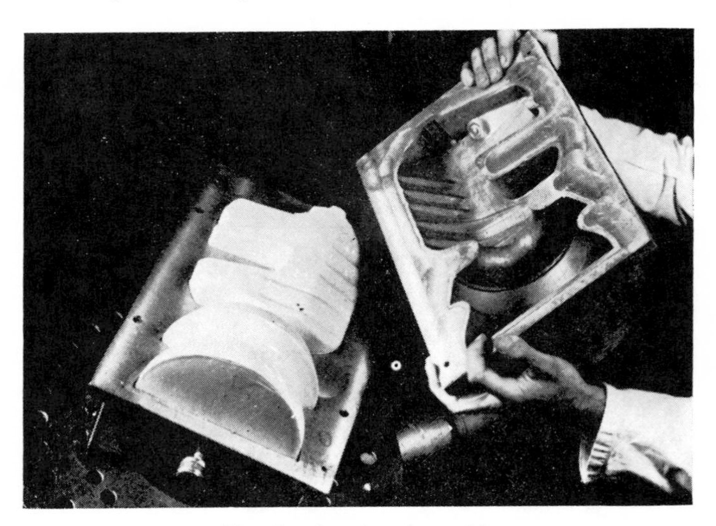

Fig. 5. Opening the mold.

Wall Thickness. The type of equipment used for rotational molding determines the wall thickness of the mold. Molds to be used in a hot-liquid machine should have a greater wall thickness (about $\frac{1}{2}$ in.) than molds to be used in a hot-air convection oven, which should be about ¼-in. thick. Thicker mold walls permit the high temperatures caused by the fast heat transfer of the liquid to heat the powder uniformly

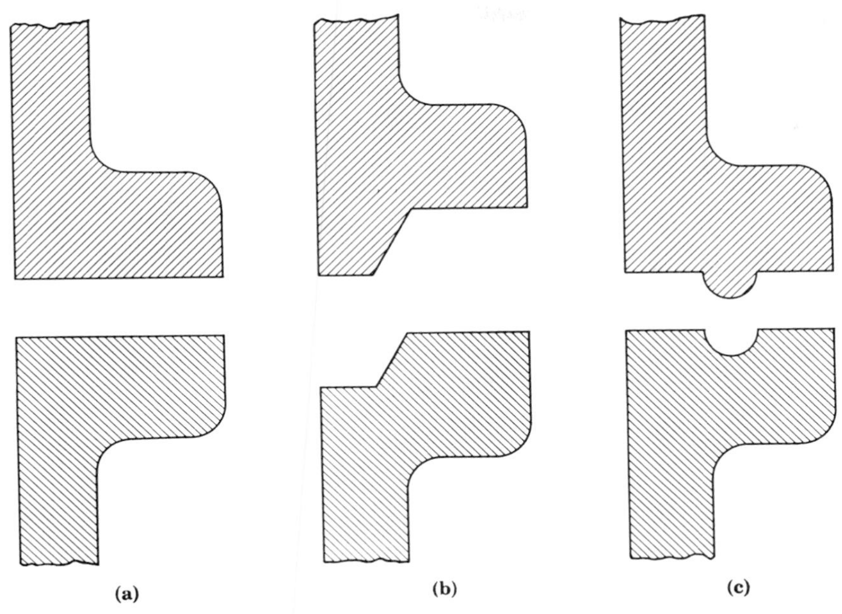

Fig. 6. Three types of flange mating surfaces for rotational molds. (**a**) Flat; (**b**) offset; and (**c**) tongue-and-groove.

without causing hot spots and with less risk of mold warpage. Thinner mold walls will result in a reduction of cycle time in a convection oven. Steel molds should be constructed of 10–18-gage (3.3–1.2 mm) steel, whenever possible. They should have flat steel flanges which should be stress relieved and machine matched after being welded to the mold.

Flange Mating Surfaces and Hinges. Any of the three types of flange mating surfaces shown in Figure 6 are suitable. The flat type can be used on either convection-oven or hot-liquid molding machines. It simplifies mold maintenance but has a higher initial cost than either the offset or tongue-and-groove surface. The mating surfaces should be machined smooth for a good fit, and the molds should be stress relieved before the parting lines are matched. Hinges should be located away from the parting line so that the mold halves are allowed to come together freely and subsequent clamping pressure can be more equally distributed.

Insulating Lids and Coverings. When openings are desired in rotationally molded pieces, insulating lids or inserts can be used. Any type of insulating material can be used which keeps the powder from fusing at the area over which it is applied. Fluorocarbon polymers (Teflon, Du Pont), asbestos board, and silicon foams are commonly used. If thin-wall sections are desired in a molded piece, they may be obtained by covering a section of the mold with an insulating material which will result in a smaller amount of material sticking to it at that point. When infrared mold heating is used, wall thickness may be controlled by painting. Dark paints on the outer mold surface will result in higher heat absorption and consequently in thicker walls in the molded part. Light-colored paints, or aluminum paint, will reflect the heat and result in thinner walls. Paints used for this purpose must be thermally stable at the high oven temperatures used. Paints have a relatively small effect on wall thickness compared with insulating covers made of materials such as asbestos.

Venting. Venting of rotational molds is often recommended to maintain atmospheric pressure inside the closed mold during the entire molding cycle. A vent will reduce flash, as well as piece or mold distortion, and will lower the pressure needed to keep the mold closed. It will also prevent blowouts caused by pressure. Finally, it will permit the use of thinner-walled molds. Figure 7 is a schematic presentation of a vented mold. The vent is a thin-walled metal or Teflon tube which extends to near the center of the mold. It must enter the mold at a point where the opening which it will leave will not harm the appearance or utility of the molded item. The vent is filled with a material, such as glass wool, to keep the powder charge from entering the vent during rotation. The end of the vent outside the mold should be protected so that no water will enter during cooling, or so that hot liquid will not enter during heating.

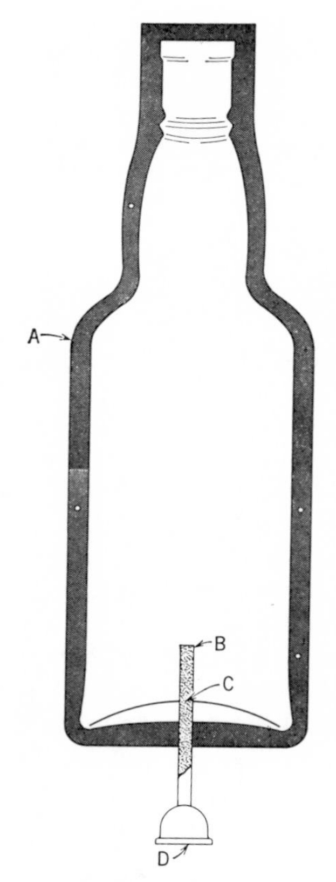

Fig. 7. Diagram of a vented rotational mold for a display bottle. Key: A, mold; B, tube of stainless steel or fluorocarbon polymer (Teflon, Du Pont); C, glass wool; D, oil filter cap.

Mold Release. To obtain satisfactory mold release of powdered polyethylene parts, it is recommended that the mold cavities be kept clean and that a proper mold-release agent be used. Most foreign material found in mold cavities can be removed with an abrasive cloth such as emery paper. Molds which have previously been used with vinyl resin or other plastic material may require special cleaning. The most effective way to remove such material from the mold surface is sandblasting; however, aluminum oxide rather than sand should be used. This procedure will result in a matte-finished surface. Once the cavities are clean, one of the semipermanent fluid mold-release agents can be applied to aid in part removal. If properly applied and heat cured, these release agents will not adversely affect later paint adhesion after flame treatment.

Generally, the rotational molding of vinyl plastisols does not require the use of a release agent.

Mold-Handling Systems

The mold-handling system includes all the equipment required to transport the mold (or molds) through the loading, heating, cooling, and unloading steps and to bring about the biaxial rotation typical of rotational molding. The main parts of the system are the spindle, the mold mount (spider), and the drive (Fig. 8). A machine can have one or several spindles. With the single-spindle machine, such as the one shown in Figure 8, the production cycle time is the sum of the intervals required to complete all the processing steps, ie, mold charging, heating, cooling, and part removal. The single-spindle machine is suitable for molding prototype pieces and for small production runs.

By using a multiple-spindle, or carousel-type, machine (Fig. 9) production rates can be increased. Multiple-spindle machines are usually sheel shaped. The spindles, each carrying a group of molds on a single large mold, are mounted on a common hub. The wheel is indexed horizontally from station to station.

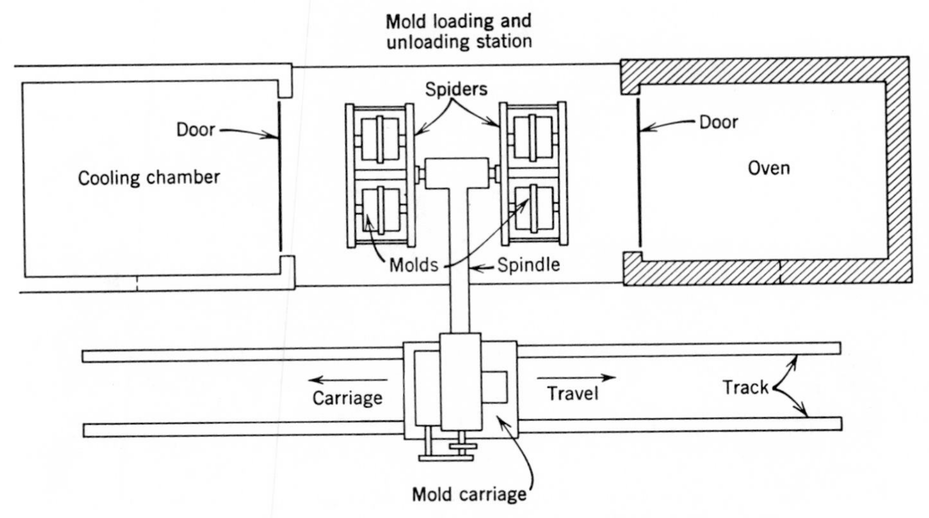

Fig. 8. Diagram of a single-spindle rotational-molding machine.

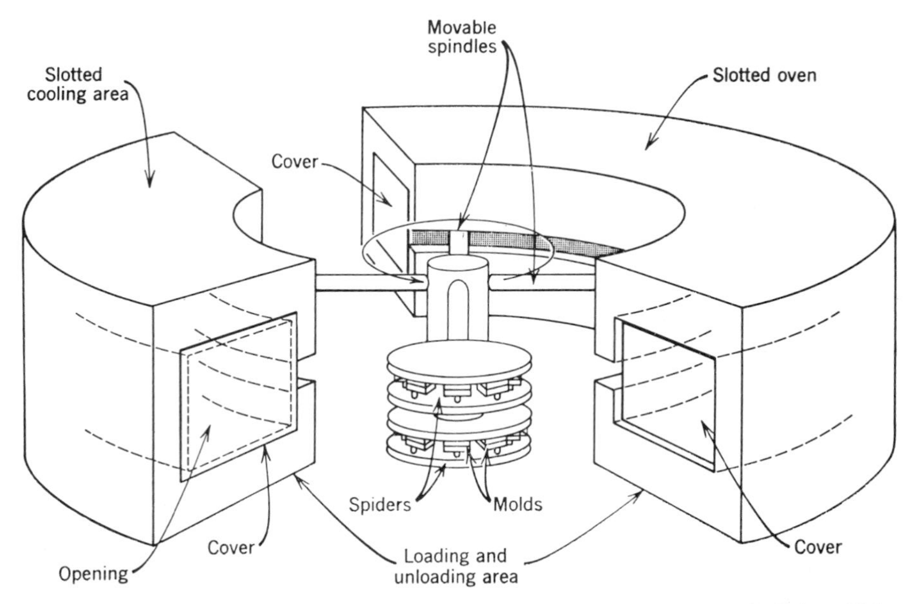

Fig. 9. Diagram of a typical rotary, multiple-spindle (in this case a four-spindle) mold-handling system.

Figure 10 shows a comparable three-spindle rotational-molding machine. The spindle with the multicavity molds at left is about to enter the oven. The spindle with the large single-cavity mold is just emerging from the cooling station at the right. The spiders with the molds are moved clockwise, first through the gas-heated oven and

Fig. 10. A three-spindle rotational-molding machine. Courtesy McNeil Corp.

then through the water-spray cooling unit in a circular path. The three-station unit is a much-used multiple-spindle machine. It is available with either hot-air or liquid spray heating. While one spindle is in the oven, a second spindle is being cooled. In the third station, both part removal and resin loading take place.

If one spindle and one station of a rotary unit are used for each operation, production time will be roughly that needed to complete the longest processing step. Frequently, two operations, for instance piece removal and mold loading, can be combined into one station, thus reducing the number of spindles required. However, the addition of more spindles permits more time for processing steps which take longer. This increases production rates. For instance, the molding unit can be designed so that at least two spindles are in the heating unit at all times. This permits more time for heating if this is the longest part of the cycle, which is usually the case.

Mold-handling systems vary greatly in the way they are used. In the design schematically shown in Figure 8, the spindle and drive are mounted on a carriage, which moves on tracks between three positions. Following the heating step, the carriage and molds are moved to the cooling chamber and then to the loading and unloading station.

In the pivoting-arm design, the spindle is attached to a pivot instead of being mounted on a carriage. The arm swings in a 90° arc when moving from the oven to the cooling chamber (Fig. 11). As indicated in the figure, a second pivot spindle and cooling chamber can be added. This doubles the production rate without requiring another oven.

To reduce tooling costs for low-volume production, different parts may be run on each spindle. If each part requires similar heating or cooling periods, only individual speed controls for each spindle are required. If heating or cooling periods differ markedly between parts, individual cycle controls for each spindle are also necessary. This reduces efficiency.

In the fixed-spindle rotational-molding machine (Fig. 12), which has been in use for a number of years in vinyl plastisol molding, a roller-top table is used to transfer the mold assembly between the three processing stations. Two drive systems, one in the heating chamber and another in the cooling chamber, are required to rotate the molds.

Figure 13 illustrates two ways of mounting molds on the spindle. The most common method is to attach the mold (or several molds) to platforms (spiders) lying directly above and below the spindle (double centerline mounting). This mounting is preferred for the production of small items because it permits easier loading and unloading of the mold. For molding large items, the so-called offset arm is preferable. This way of mold mounting makes possible the rotation of much larger parts. Moreover, it minimizes variations in wall thickness by bringing the center of volume of the mold close to the intersection of the primary and secondary axes of rotation. A wide difference in peripheral velocity from top to bottom, caused by a greatly offset mold, may result in poor material distribution and nonuniform wall thickness of the moldings.

Drive Systems for Biaxial Mold Rotation. A variety of drive systems is available to provide biaxial rotation of the molds. They include either a variable-speed drive or constant-speed drive with interchangeable gears or chain sprockets.

Most rotational molding is carried out with a 10- to 12-rpm rotation of the primary axis and a 2- to 3-rpm rotation of the secondary axis, that is, an average ratio of

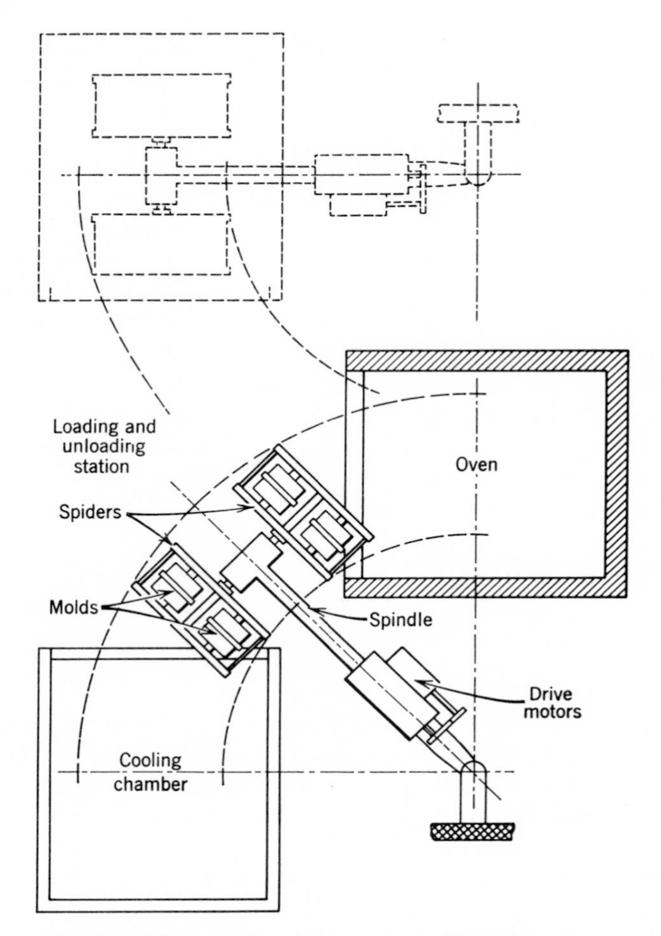

Fig. 11. Diagram of a pivoting-arm mold-moving system.

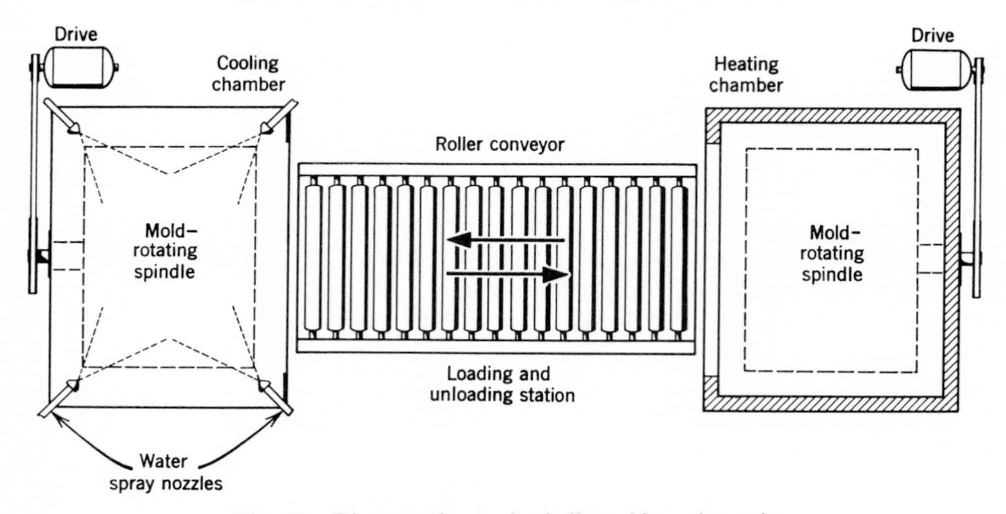

Fig. 12. Diagram of a fixed-spindle mold-moving unit.

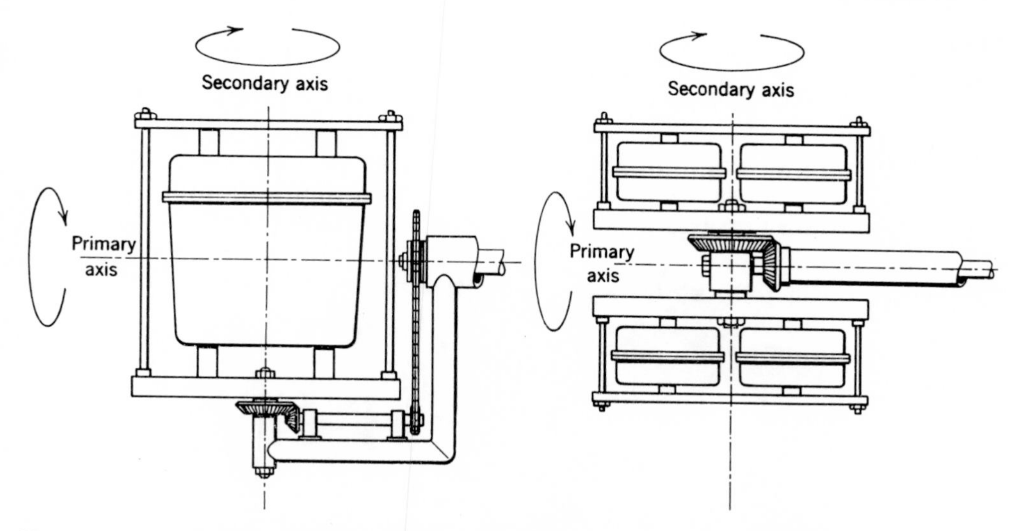

Fig. 13. Mold-mounting systems. Right: double centerline mounting. Left: Offset-arm mounting.

4 or 5 to 1. This ratio has proved satisfactory for many applications. However, provisions should be made for changing the ratio to correct nonuniform powder distribution and wall thickness which can occur at the average rotation ratio. If many different parts are to be produced on one machine, variable-speed drives are recommended to eliminate the need to change gears or sprockets when changing speed or ratio of rotation. If the drive allows for variable-speed rotation of each axis independently over a range of 1 to 25 rpm, any practical rotational ratio can be obtained.

Mold-Heating Systems

The three most commonly used methods by which molds are heated are hot-air convection, hot-liquid conduction, and infrared radiation. Either gas-fired or electrical heaters may be used. The most suitable type of heat source depends mainly on the geometry of the piece, the number of molds on each spider, and the required production rate.

Normal oven temperatures are 400–700°F (205–370°C) and frequently as high as 900°F (480°C). This high temperature will usually not damage thin-walled moldings, but it may damage some molds. Ovens must be well insulated to minimize heat losses.

Hot-Air Convection Heating. Hot air is the most commonly used heat source for rotational molding (Fig. 10). This heat source is particularly suitable for molding thin-walled pieces (less than 60 mil (1.5 mm)). Oven temperatures may vary from 400 to 900°F (205–480°C). The higher the oven temperature, in this range, the shorter the heating cycle.

Accurately directed air flow is important. This can be obtained by baffling the air blower. Equally important are the velocity of the air and uniformity of the temperature throughout the oven. A high-velocity scrubbing action is needed for maximum heat transfer. Hot spots in the oven must be avoided when items of uniform wall thickness are to be molded. Excess heat capacity of the oven is desirable so that the required operating temperature can be rapidly attained after insertion of the cool, charged mold.

Gas-fired, oil-fired, or electrically heated hot-air ovens are available. If temperatures are frequently changed, a recording temperature controller is recommended.

Hot-Liquid Conduction Heating. Heating by means of hot-fluid sprays is also much in use. In such ovens, a noncorrosive, eutectic mixture of inorganic salts commonly used for heat treating of metals is the heat transfer medium; solvents are

Fig. 14. A hot-liquid oven. Courtesy E. B. Blue Co.

not used in these sprays. Molten wax and high-temperature heat-transfer oils have been used, though less successfully. Figure 14 shows a hot-salt spraying oven.

Operating temperatures in the liquid-spray system are normally in the range of 450–550°F (230–290°C). The heat-transfer liquid is brought up to operating temperature in a melting tank, or reservoir, containing gas fire tubes or electrical immersion heaters. A heating period of up to 2½ hours, depending upon the volume of heating medium used, is required to reach operating temperature. The fluid is pumped through a network of spray nozzles onto the rotating mold assembly. It is important that the melting tanks have enough capacity to maintain the operating temperature throughout the heating cycle. The temperature drop in the melting tank should not exceed 10–15°F (5.5–8.5°C) during the mold-heating cycle.

In this heating system, heat is transferred by conduction rather than convection. Hot-liquid spraying, therefore, normally heats the molds faster than hot air. It is generally better suited for molding pieces with heavy walls and for the production of more complex shapes.

Before leaving the oven the molds are usually rinsed with water to remove the heating fluid. Since unrecoverable heating fluid can mean the difference between profit and loss in hot-liquid spray heating, it is essential that such equipment have a good system to recover and reuse the heating medium.

Infrared Heating. Infrared heating with either gas-fired or electrical radiation heaters is mainly used on special-purpose and prototype rotational-molding equipment.

Heating with infrared is very fast and efficient, but this heat source is limited to molding simple shapes and single molds. Multiple or complex-shaped molds cannot be heated by infrared because the radiant heat cannot strike all parts of the mold uniformly. Shielded areas of the mold will not attain as high a temperature as exposed parts of the mold surface. Consequently, thin spots will occur in the wall of the molded item.

Mold Cooling

Spraying with water is the most common method of cooling the mold. During the cooling process the mold should be rotated; failure to do so will result in the soft molten polymer sagging toward the bottom of the mold. A highly atomized water spray results in uniform and rapid cooling. Blowers or compressed-air nozzles in the cooling chamber increase flexibility of the cooling system. For critical cooling applications, air and water can be used intermittently.

Since quick cooling makes the overall molding cycle shorter, it is desirable for economic reasons. However, if cooling is too rapid, warpage of the molded piece may occur. If this happens, the cooling rate should be slowed by raising the temperature of the cooling bath or by using forced air in combination with the water spray.

Mold Filling

Molds may be filled manually or automatically. For manual filling, the required quantities of powder are premeasured, by weight or volume, in a container and then dumped into the lower halves of the molds. Manual filling is advantageous when filling differently colored items simultaneously in a multiple-cavity mold.

For long runs, automatic mold-filling equipment is useful. The bulk density, measured in lb/ft^3 (1 lb/ft^3 = 16 g/l), of powdered polyolefins may vary, and to avoid differences in wall thickness between pieces, the powder charge should be based on a weight measurement. However, volumetric dispensers are also used for powdered polyolefins, and are readily adaptable for measuring vinyl plastisols in rotational casting.

Internal Mold Blanketing

Equipment through which inert gas and cooling media can be introduced into and vented from the mold during the heating and cooling cycles is recommended for rotational molding (Fig. 15). A system of this type will prove particularly useful in applications where comparatively high molding temperatures are necessary, such as when rotationally molding high-density, or linear, polyethylene. Such systems improve the properties of the end product, reduce odor and decoloration of the molded item, and shorten cooling time.

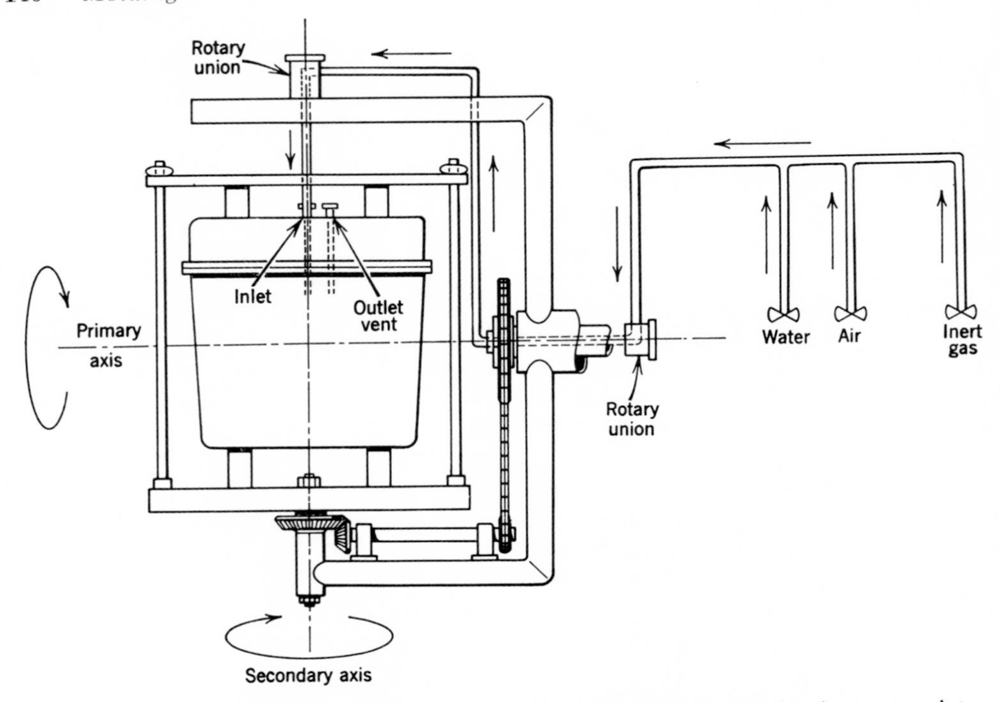

Fig. 15. Diagram of typical auxiliary equipment for introducing gas, cool moist air, or water into a rotational mold.

Molding takes place in the presence of air, with its oxygen content, entrapped in the mold cavity. The introduction of an inert gas such as nitrogen or carbon dioxide into the mold during the heating cycle flushes out the oxygen and thus prevents oxidative degradation of the resin while it is being fused. Such resin deterioration may occur if the resin is heated excessively. Thus, flushing out of the air will eliminate undesirable odor and discoloration in the interior of the molding, while simultaneously improving its physical properties.

Once the molded piece enters the cooling cycle, cool air can be substituted for the inert gas. This can be followed by the introduction of moist air and finally water to help cool the piece from the inside. Faster and more uniform cooling made possible by this technique reduces cycling time, decreases the crystalline content in the molded piece and thus improves its toughness, and controls warpage due to simultaneously cooling the hollow molding from both its interior and exterior.

The inert gas and cooling media are introduced into the mold through hollow rotating drive shafts, mechanical seals, and a flexible high-temperature hose inserted in the mold (Fig. 15). To make the most efficient use of such auxiliary mold-flushing equipment, the following steps are observed: (a) Enough inert gas is introduced into the mold to flush out all the oxygen. Then the gas volume is reduced just enough to keep oxygen from reentering through the outlet vent. (b) Shortly after the cooling cycle begins, the gas is shut off and air is introduced into the mold. (c) Once the interior layer of the item is cooled (the period required will vary with the size and shape of the moldings but should be ended shortly after the air is turned on) a small amount of water can be added to speed cooling. Then water input can be slowly increased until the piece is sufficiently cooled for removal from the mold.

Applications

Rotational molding of polyolefin powders has grown significantly since its inception, and the marketing outlook indicates that the once specialty process is rapidly emerging as one of the more versatile approaches to the manufacturing of a broad range of products. The technique has already been accepted as particularly suitable for applications such as display-window mannequins and other display figures of all sizes, sporting equipment, complex-shaped toys, and other items for children such as children's furniture.

More recently, applications of rotational molding, some still in the development stage, are in battery cases, traffic cones, luggage, housings of various kinds (Fig. 16);

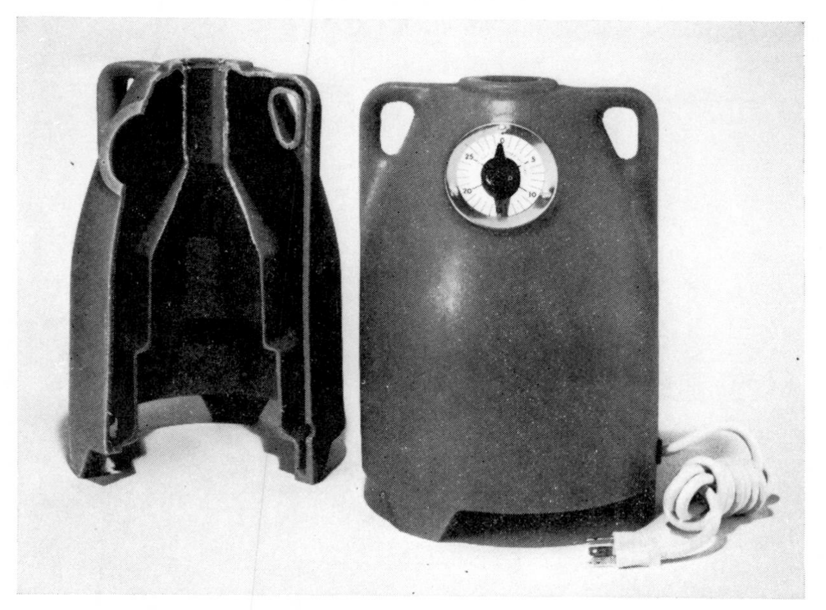

Fig. 16. A double-walled hydromassage motor-fan housing, illustrating complex features than can be molded by rotational-molding techniques. Courtesy C/R Custom Plastics Corp.

waste receptacles, water tanks, and automotive parts including air ducts, fuel tanks, glove boxes, and fender liners. The market prospects for these items alone represent hundreds of million of pounds, a very desirable volume.

Economic Aspects

When studying the economics of rotational molding, the most logical approach is to compare it to the other molding techniques with which it is competing: blow molding and injection molding. However, when comparing the cost per piece (unit cost) of the three molding processes, it must be kept in mind that these techniques have been designed for different purposes. Injection molding yields high-precision items at high production rates, but generally cannot be used to make one-piece closed vessels. Blow molding is best suited for making bottles and similarly shaped items which require polyolefin resins with low melt indexes. Rotational molding is best, although by far not exclusively, suited for producing closed, thick-walled items.

The low cost of rotational molds compared with blow molds and, especially, injection molds is estimated to amount to about 90% savings. Moreover, mold lead time is about one month as compared with four to five months for injection and blow molds.

Changes in resin melt index and density affect the cycle time in rotational molding more than in injection molding and blow molding. However, increases in wall thickness affect the cycle time much less than in injection or blow molding. The wall thickness of a rotationally molded item can be increased merely by adding additional powder to the mold; increase of wall thickness in injection molding, on the other hand, requires a complete modification of the mold. Also, design alterations are usually made much more easily and at much less cost in rotational molding.

Intensive comparative-cost studies for containers of sizes varying from 1 to 250 gal have shown that at low production levels (10,000 units yearly), rotational molding

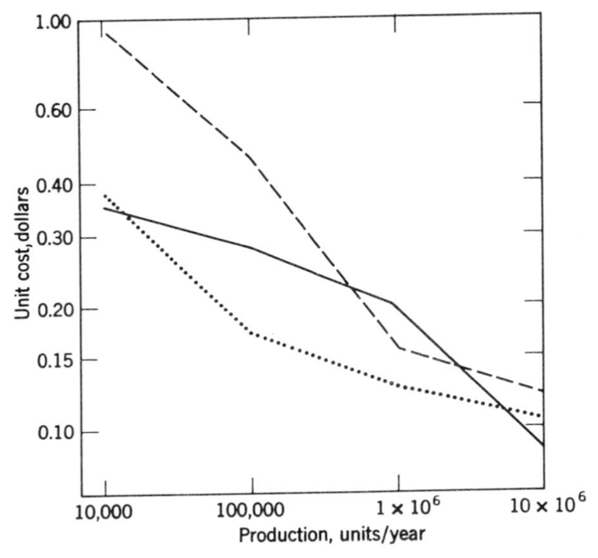

Fig. 17. Unit costs, 1-gal containers. Legend: solid line, rotational molding; dashed line, injection molding; dotted line, blow molding.

had the lowest unit cost for all sizes, injection molding the highest. This indicates that for short-run and prototype moldings, rotational molding is economically superior to injection or blow molding, provided the item is suited for rotational molding.

Figure 17 shows that at low production rates, 1-gal containers are blown almost as cheaply as they are rotationally molded, whereas injection molding costs three times as much. At higher production rates, eg, 100,000 and 1,000,000 units annually, blow molding costs become lower than costs for rotational and injection molding. But at the production rate of 10,000,000 units, unit costs revert to the order characteristic for low production rates; rotational molding is cheapest, injection molding most expensive, and blow molding in between.

The comparative costs for 5-gal and 20-gal containers show little difference in unit cost for all three molding techniques, especially at very high production rates (10,000,000 units for the 5-gal containers and 1,000,000 units for the 20-gal containers).

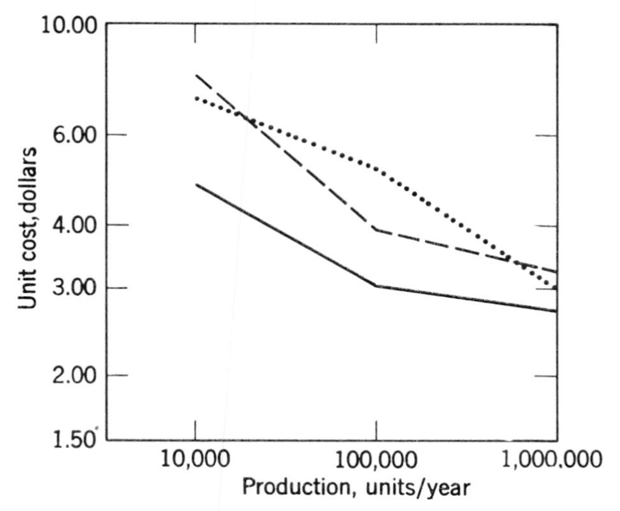

Fig. 18. Unit costs, 55-gal containers. Legend: solid line, rotational molding; dashed line, injection molding; dotted line, blow molding.

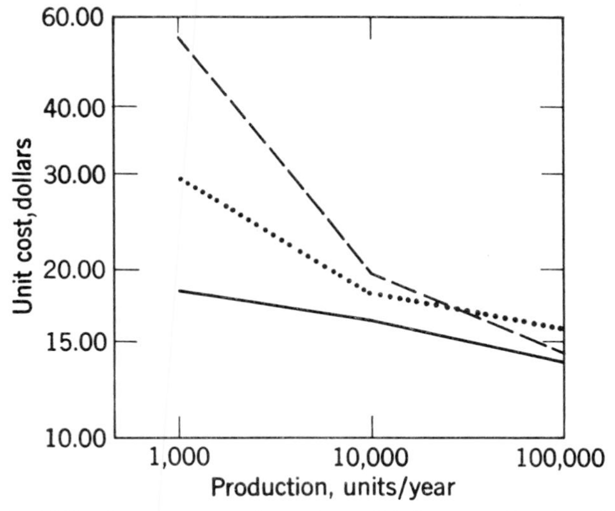

Fig. 19. Unit costs, 250-gal containers. Legend: solid line, rotational molding; dashed line, injection molding; dotted line, blow molding.

For 55-gal containers, the economics of rotational molding are markedly superior at any annual production rate, from 10,000 to 1,000,000 units (Fig. 18). For a still larger container, a 250-gal tote bin, rotational molding is by far the most economical molding technique, especially at the low production rates (Fig. 19).

For all three molding techniques and for all sizes of containers or other end products, unit costs at any annual production rate depends on a combination of factors of varying influences, mainly investment in and amortization of equipment and molds, labor and operating expenses, raw materials, and taxes. The study on 1-gal containers discussed above, for instance, implied a change from low-production to high-production rotational-molding equipment when annual production exceeded 1,000,000 units. Without such a change in equipment, rotational molding would be uneconomical at the 1,000,000-plus annual production rate.

Advances in improved equipment and technology have progressed to the point where rotational molding of polyolefin powders is beginning to challenge the other molding techniques as a major processing system. Simultaneously, resin development is broadening the spectrum of powdered materials to meet the demands of the industry. This is not to indicate that the development of the technique is complete, for more sophisticated equipment, improved heat control, less manual labor required in loading and unloading, and faster cycles are needed for high production rates. The rotational-molding industry is confident of the capabilities of the technique and anticipate substantially larger markets in the near future.

Bibliography

"New Dimensions in Rotomolding," *Mod. Plastics* **43** (April 1966).

"Plastisol Molding," *Modern Plastics Encyclopedia*, Vol. 41, No. 1A, Breskin Publications, Inc., New York, 1964, pp. 676–682.

Rotational Molding of Microthene Polyethylene Powders, Booklet No. PL 23-865, U.S. Industrial Chemicals Co.

W. C. Johnson, *Economic Evaluation of Rotational Molding, Symposium on Rotational Molding, U.S. Industrial Chemicals Co., Nov. 1963.*

Richard E. Duncan, David R. Ellis, and Robert A. McCord
U.S. Industrial Chemicals Company

AUXILIARY PROCEDURES

Auxiliary procedures associated with molding of plastics are of importance for their effects on both the quality and the economics of the molded article. An unsatisfactory product or excessive cost may result from the failure to consider these essential adjuncts to the actual molding process. This article discusses some of the more important procedures, including those related to the pretreatment and handling of the raw materials, those associated directly with the molding operation, and those required as secondary, postmolding processes.

For purposes of this discussion, the auxiliary procedures will be oriented principally toward injection molding. However, it should be understood that many of these procedures are broadly applicable to other plastics fabrication techniques, such as compression or transfer molding, blow molding, and extrusion. See other sections of MOLDING article.

Handling of Raw Materials

Plastics for molding operations are usually furnished in pellets approximately ⅛ × ⅛ in. made either by dicing sheet material or by chopping extruded rods. Suppliers package the pellets in 50-lb bags, 300-lb drums, or 1000-lb cartons on disposable pallets. For large users, materials can be furnished in bulk shipments in special truck bodies or rail cars. The choice of handling system is based on the economics of each case. A producer who uses a variety of materials in small, close-tolerance parts would not use the same handling system as a producer of a large volume of large parts in a single material.

The simplest way of handling material is to use a scoop and a 50-lb bag. The 50-lb bags can be carried or wheeled to the press on a small two-wheeled hand truck. Even 300-lb drums can be handled in this manner, but a fork truck and pallets are more economical if many drums are to be moved. When there is greater demand for transferring material to larger presses with higher material hoppers, it becomes advantageous to utilize some form of automatic handling such as a pneumatic or vacuum loading system.

Pneumatic Loading Systems. A pneumatic loading system, such as that shown in Figure 1, uses compressed air and a venturi tube which draws the material from the drum or carton and blows it into the press hopper. This loader can be equipped with controls for automatic operation and thus can keep the machine hopper

Fig. 1. Pneumatic loader for loading from a drum beside the press. Compressed air and electric service are required. Courtesy Whitlock Associates, Inc.

filled without attention from the operator or material handler. On a high machine that requires filling several times a day this feature can be particularly worthwhile when lost time, inconvenience, and startup losses due to running out of material are considered. The machine can also be furnished with removable pickup tubes to allow easy drum replacement and with rotating pickup tubes to help feed irregularly shaped material such as regrind.

Pneumatic loading systems have capacities from a few hundred pounds per hour to about 2000 pounds per hour, depending upon the length of run, amount of lift, and continuity of operation. Disadvantages of the pneumatic system are the relatively high cost of compressed air, the danger of material contamination if compressed air is not clean and dry, and the housekeeping problems if the material is dusty. (The granulating operation produces some dust in most plastics.) Because the pneumatic system draws material to the venturi and blows it from the venturi tube to the hopper, a filtering system must be devised if dust in the air around the machine is objectionable.

Vacuum Loading Systems. The vacuum system uses a vacuum pump driven by an electric motor to generate a vacuum which draws the material all the way to the hopper. This places the entire system under a negative pressure and necessitates a filter to prevent fines from returning to the vacuum pump.

The simplest form of vacuum system has an integral vacuum pump, motor, and filter mounted on top of a vacuum chamber, as shown in Figure 2. The entire unit is fastened to the top of the machine hopper. A hopper level-control switch in the molding-machine hopper turns on the vacuum motor, and the initial rush of air closes a rubber valve at the bottom of the vacuum chamber. The resulting vacuum causes air to flow through the feed tube, drawing pellets from the material drum into the vacuum chamber. The vacuum pump pulls the air through the filter for a predetermined time and then stops. It shakes the filter or reverses and blows air through it to clean off all the dust, thus adding it back to the load of material from which it came. The weight of the pellets opens the rubber valve at the bottom of the vacuum chamber

Fig. 2. Vacuum hopper showing electric motor at top and rubber valve at bottom. The whole unit sits on top of machine hopper. Courtesy Conair, Inc.

and the pellets drop into the machine hopper. This cycle is repeated until the level-control switch is operated, indicating that the desired level has been reached. (Figure 3 shows a vacuum system with a cyclone separator for very dusty material.)

The vacuum system is flexible in that it can be adapted to a wide range of material-handling needs. Figure 4 shows a central vacuum pump capable of feeding two

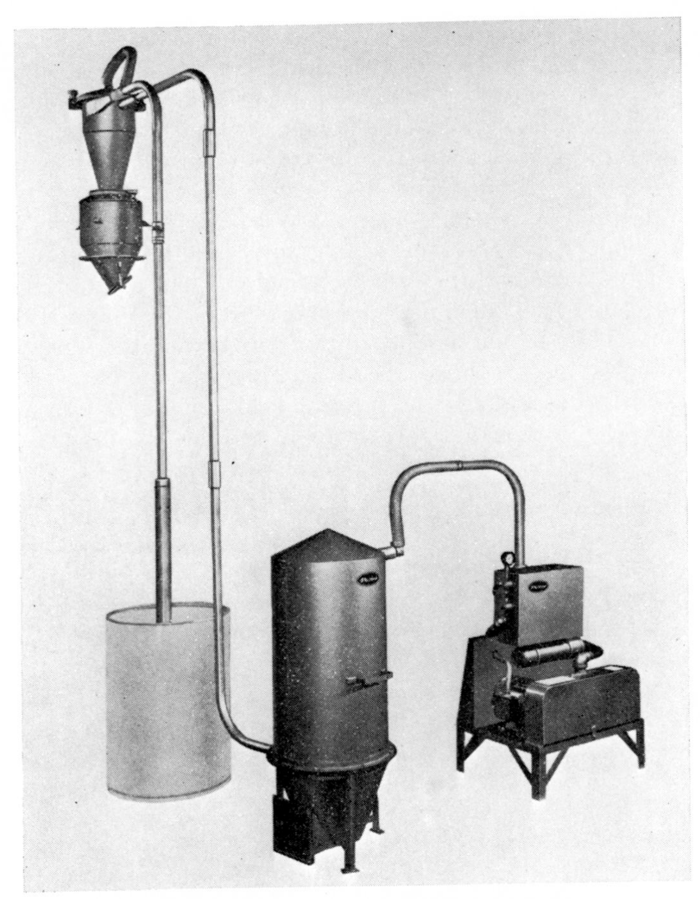

Fig. 3. Vacuum system showing use of cyclone separator for very dusty materials. Courtesy Whitlock Associates, Inc.

materials to the same press. These could be virgin material and reground material; they could also be clear material and color-concentrate material.

If a number of presses are to be equipped with vacuum hoppers, it is possible to use a central vacuum pump with a sequencing switch which will initiate checking and filling of each machine hopper at a selected interval. This reduces the investment in vacuum pumps, but the reduction must be balanced against the increased cost of vacuum lines. Figure 5 is a schematic diagram illustrating how one material can be fed to a number of machines. Figure 6 shows how one vacuum power unit can be used to feed different materials to a number of machines.

Figure 7 shows the unloading of a rail car or truck by means of a vacuum system into boxes or bins inside the plant. Figure 8 depicts a relatively complicated system by which a rail car can be unloaded into any one of four silos, each of which can be fed to any one of eight presses that are using different types or different colors of material. This system combines great flexibility with the use of very little horsepower. The horsepower required to move plastic pellets for various distances and heights can be obtained from the chart in Figure 9.

This chart is representative of the conveying rates for free-flowing pelletized materials with the various sizes of power units. Although the rates shown are based on controlled test conditions with a virgin polyethylene material, the size of unit needed for an installation can be approximated by determining the pounds per hour needed and the vertical and horizontal distances. (Note: One foot of vertical distance equals two feet of horizontal distance.) Different types and shapes of materials, the number of bends in the material lines, and the amount of regrind to be conveyed can affect the conveying rates.

Couplings are made to close tolerances to fit the aluminum tubing material line so that there are no ledges or crevices where granules can lodge. O-rings are used in the fittings to prevent air leakage. The system should be grounded to prevent the flow of plastic from generating a static charge that would cause a buildup of dust. Otherwise, letting the vacuum pull a clean rag through the system usually suffices to remove the remaining dust when changing colors or materials. Consideration must be given to the air supply, since unloading a rail car with warm humid air might overload subsequent drying operations on moisture-sensitive materials. Contamination of the

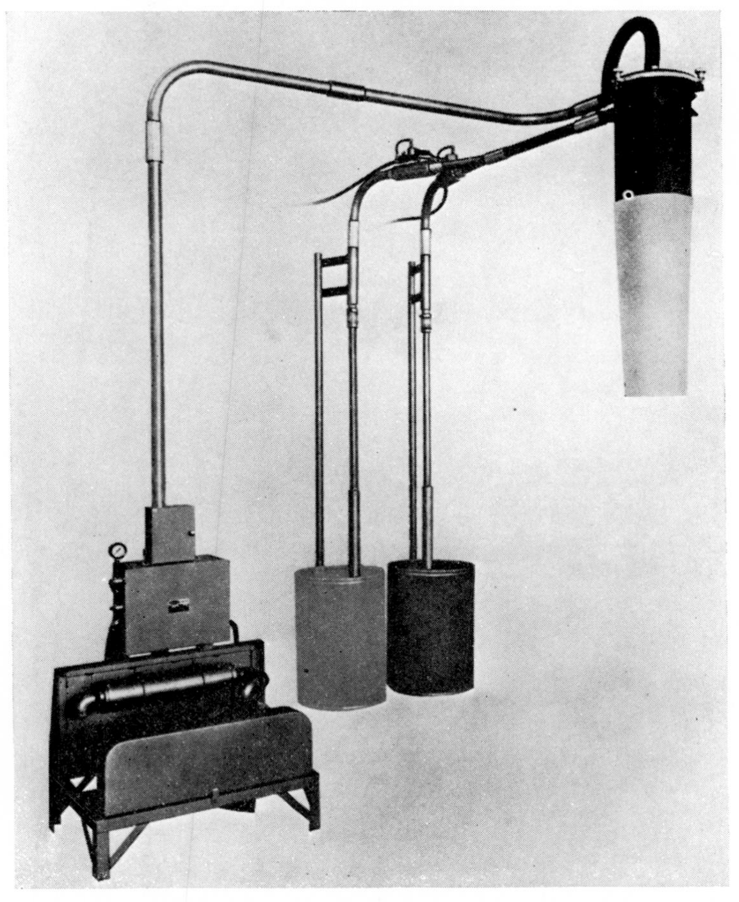

Fig. 4. Central vacuum pump feeding two materials to one (machine) hopper. Vacuum pump with sequencing switch can serve several hoppers. Courtesy Whitlock Associates, Inc.

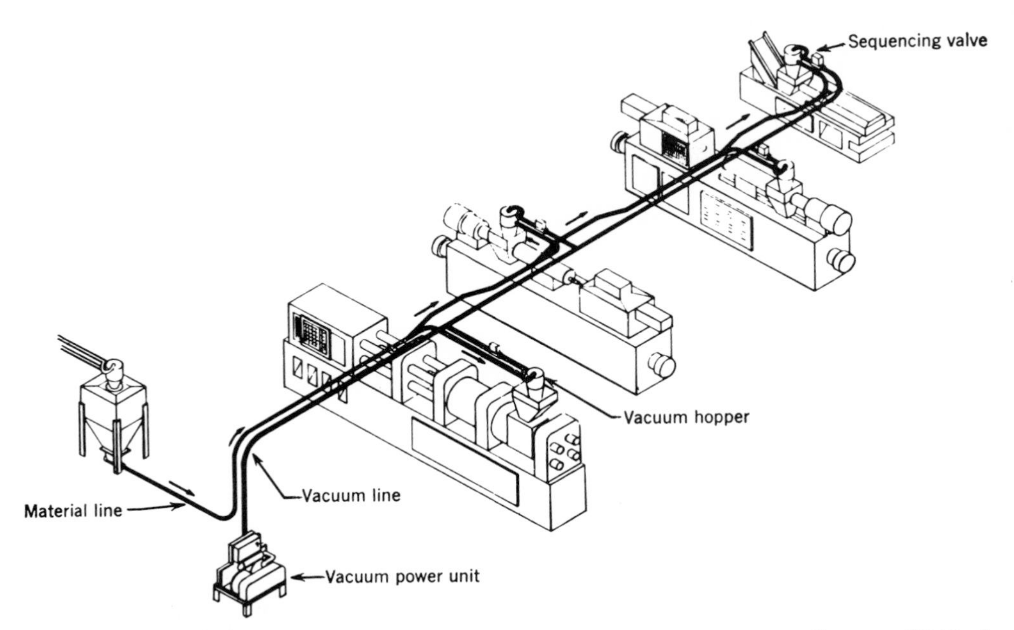

Fig. 5. Central vacuum system used to feed one material to a number of presses. Courtesy Whitlock Associates, Inc.

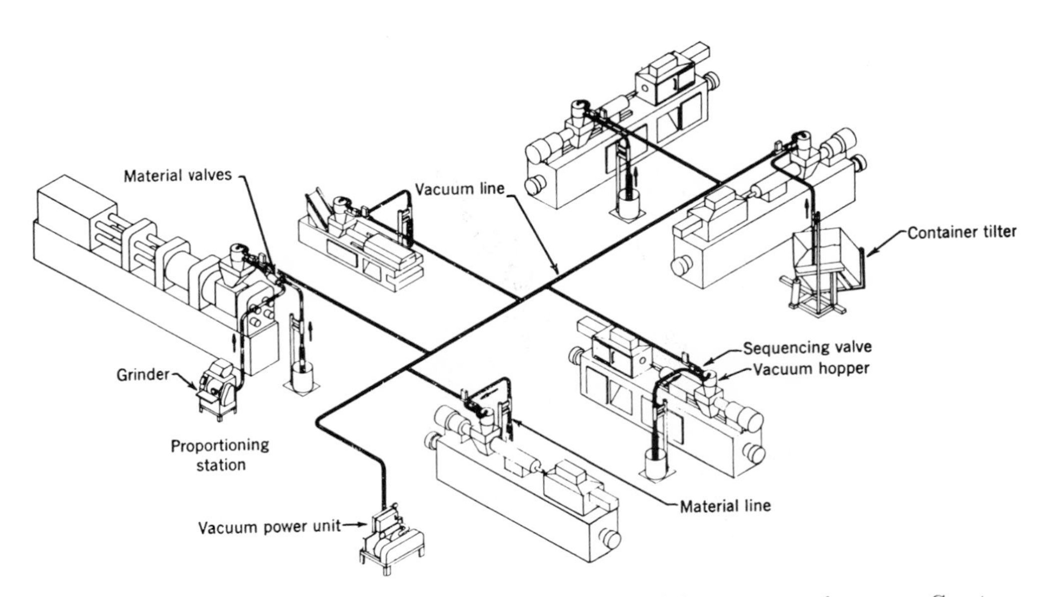

Fig. 6. Central vacuum system used to feed a variety of materials to a group of presses. Courtesy Whitlock Associates, Inc.

material in vacuum systems has been negligible except in the case of some dry colorants that are pulverized or that have an oily consistency.

In a multiple-unit system thought must be given to the need for a spare power unit or spare motors and pumps so that the whole operation will not be shut down by failure of any one piece of equipment. Sometimes it is possible to operate manually and use one unit for double duty while another is being repaired.

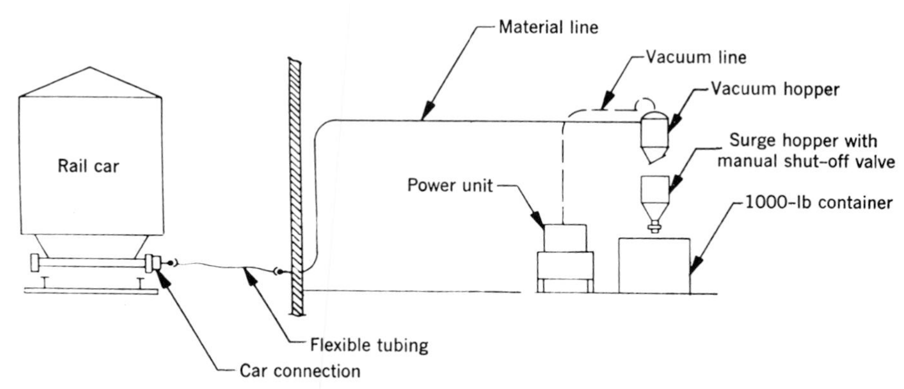

Fig. 7. Vacuum system used to unload a rail car. Courtesy Whitlock Associates, Inc.

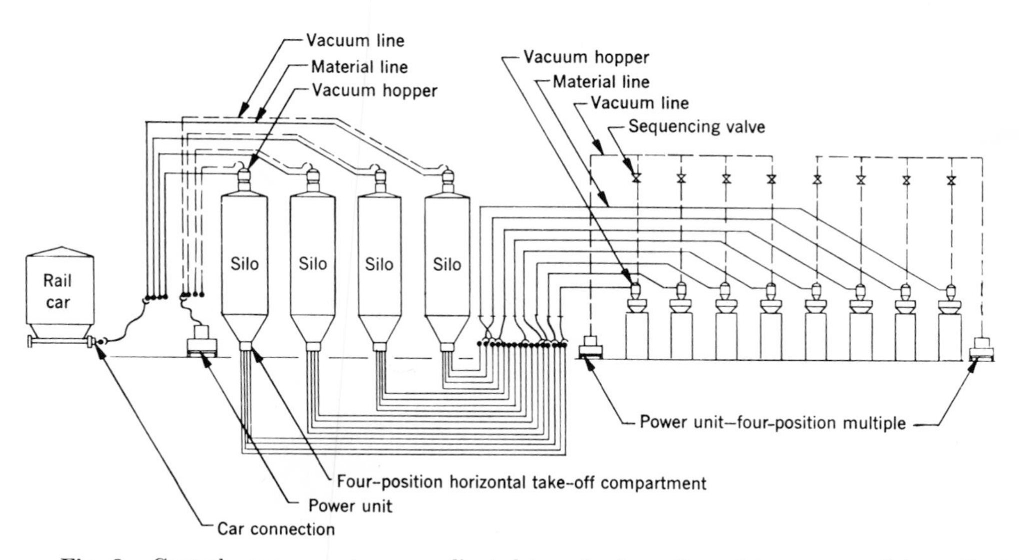

Fig. 8. Central vacuum systems coordinated to unload a rail car into any one of four silos, material from each of which can be fed to any one of eight presses. Courtesy Whitlock Associates, Inc.

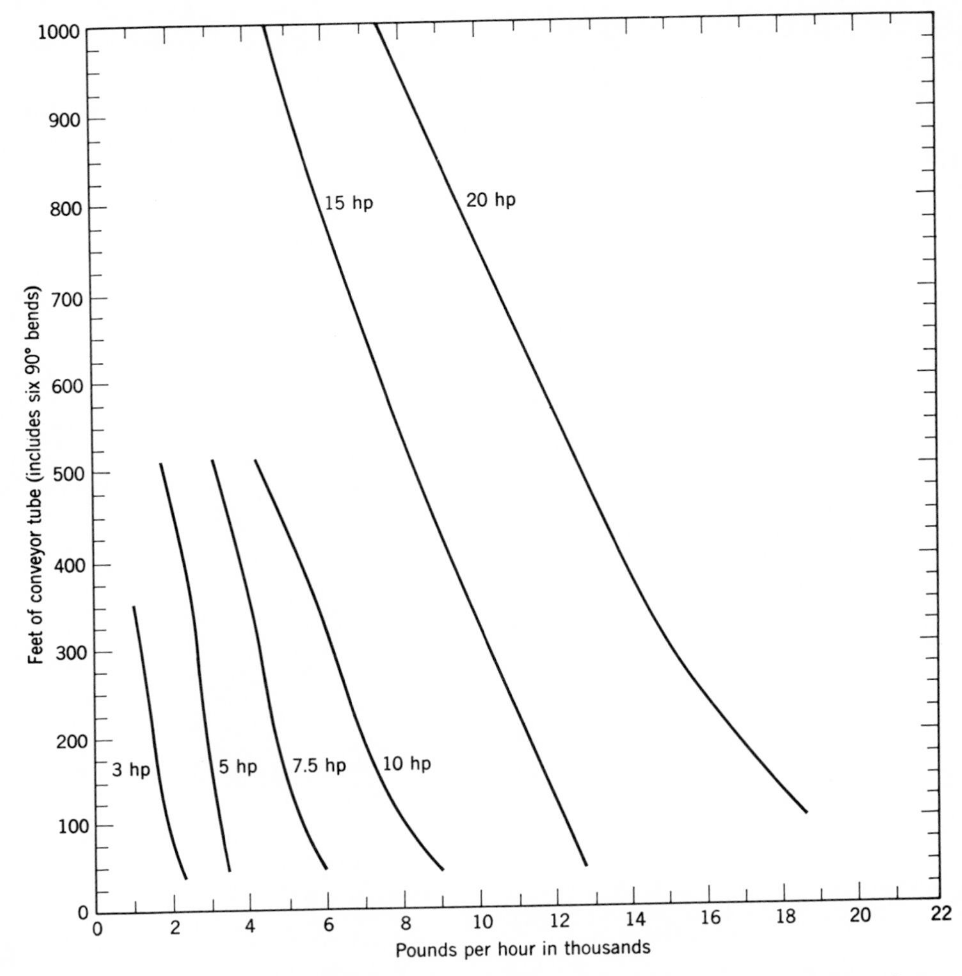

Fig. 9. Rates at which polyethylene beads are conveyed through various sizes of systems by power units indicated. (Polyethylene beads are ⅛-in. diameter and have a bulk density of 37 lb/ft³. Conveying vacuum is 12 in. mercury.) Courtesy Whitlock Associates, Inc.

Procedures Associated With the Molding Operation

Drying of Plastics. Most plastics must be dry before molding if appearance of the molded part is of any importance. Dry means anything from an 0.025% maximum moisture content for polycarbonate to an 0.25% moisture content for cellulosic plastics. Materials like acrylonitrile–butadiene–styrene plastic or polycarbonate adsorb only a surface layer of moisture which is easily removed, but the cellulosic plastics absorb moisture into the molecular structures and this is more difficult to remove. Even the physical properties of some materials are degraded if moisture is present in the heating cylinder. In some cases, the level of moisture that can be tolerated depends upon the geometry of the part.

Since there is no quick, easy, accurate method of checking moisture content of plastics, the safest course is to dry a material before molding. Some materials can be dried by blowing heated air through them; others require air that is dry even before it is heated. Here again, one must be guided by the particular material, application, and supplier's recommendations. An important side effect of hot-air drying is that the plastic granules are partially preheated when they enter the heating cylinder; this contributes to a better appearance of the molded part and may permit a shorter molding cycle.

Die Setting. All molding shops must consider the function of setting mold dies in presses and removing them. The facilities required vary widely, depending upon the size, type, and number of presses and dies. Very small dies may be set by hand but, generally speaking, at least a die setter's truck is required to transport the dies and to raise or lower them to a height where they can be slid onto a metal plate across the tie rods of the press. To set the dies for larger presses, some molding shops install a chain hoist on a monorail or a chain hoist on a cantilevered monorail that rotates about a vertical post. Portable A-frames can also be used for large molds. If an electric truck is provided for material handling, it can be used for setting dies. Some buildings designed specifically as molding shops are equipped with bridge cranes that are used for setting dies, changing cylinders, and other maintenance work.

Controlling Mold Temperature. In most molding processes, the surface temperature of the mold cavity may be a major factor in the quality and rate of production of the molded part. Depending upon the material and the part configuration, mold-temperature requirements may range from cool to very hot; temperature fluctuations must be held within limits of a few degrees. Actual surface temperatures of the cavity are difficult to measure accurately because of rapid fluctuations caused by the introduction of the hot melt. A reasonable approximation can be made by means of a surface pyrometer if it is used quickly and carefully.

Some materials give best results in hot molds; other materials give good results in warm or cool molds. A cool mold will frequently give maximum production, but at the expense of quality. Therefore, it is necessary to consider the particular material and the desired characteristics of the part before establishing the molding cycle.

Control of mold temperature is usually accomplished by circulating a controlled-temperature fluid or vapor through channels in the dies. Generally, the proper temperature of the die fluid is established experimentally using quality, appearance, maximum production rate, and minimum scrap as criteria. The location and size of the channels in the mold are critical, and the most sophisticated temperature-control methods cannot compensate for inadequacy in location and size.

For most temperature-control requirements, a commercially available, general-purpose temperature controller is applicable. In general, the upper temperature limit of these units is the boiling point of water. They usually consist of a water tank either heated electrically or cooled by water, and preferably equipped with anticipating circuits to minimize temperature fluctuations. The controller should have adequate capacity to pump water through the mold at a rate that permits only a few degrees rise in water temperature. A separate water circuit is needed for each half of the mold, because it is sometimes desirable to maintain the two halves at different temperatures.

A number of good commercial two-circuit units are available. These units should be mounted on casters and equipped with quick disconnects and rubber hoses so that they can easily be pushed out of the way for press maintenance or be replaced if trouble develops in the controller itself.

Occasionally, a part is of such design that it is not possible to provide adequate channelling in the mold, or the part is such that the fastest heat removal and shortest molding cycle are of first importance. In these cases, a water chiller must be employed because the lowest temperature that can be obtained by a general-purpose controller is that of the plant's water supply. Water chillers use a refrigerating unit to cool the water to be circulated through the mold. The refrigerating unit can be either water cooled or air cooled and, if a number of jobs need chilled water, a central chiller can furnish cool water to a general-purpose controller at each press.

Some of the newer engineering plastics such as the acetal resins and polycarbonates require hot dies, occasionally at 250°F or above. Although some general-purpose controllers are equipped with pumps and seals that can withstand these temperatures and the requisite high pressures, the possibility of a bursting hose is an ever-present safety hazard. There are commercially available controllers which use a high-temperature fluid such as oil that does not require high pressures to reach the desired temperature. These controllers are usually equipped with armored hoses to reduce the hazard. They are similar in shape and size to the general-purpose controllers.

Compression and transfer molds for phenolic resins and other thermosetting materials are frequently heated by steam at 150–200 lb pressure. This system has the advantage of being fast, uniform, and easily adjusted. Close attention must be given to putting adequate steam lines in the mold, and care must be used in connecting the steam to the mold because of the safety hazard involved. Heat transfer from the mold to the press plates should be minimized either by milling grooves in the back plates to reduce metal-to-metal contact or by using insulating material such as asbestos fiberboard. The surfaces of the insulating material must be parallel and must uniformly withstand the clamping pressure or the dies will be thrown out of parallel alignment. See also MOLDS.

Lubricating the Mold. It is generally agreed that the best mold lubrication is none at all, but this assumes that it is possible to design the mold with adequate draft and no undercuts, and that the mold is actually constructed that way. Frequently, it takes time to find and eliminate the conditions that cause sticking in a new mold. If parts are urgently required, a mold lubricant can be used until the repair can be made.

In addition to lubrication of new molds, lubricants are also used with hard-to-fill cavities that need packing, threaded parts where the path of the thread is too long to permit draft, deep draw parts, parts with shallow lettering on the sidewalls, and parts with deliberate undercuts for snap-on assembly. Two of the most widely used mold-release agents are a silicone oil and a zinc stearate powder.

Silicone mold-release agents are most often applied as a spray from an aerosol-type dispenser. Care must be used in the application, as excessive release on the mold may affect the finish of the next part to be molded; however, a little practice with the dispenser will enable the operator to deposit a fine mist on the die. Because silicone release agents are excellent antiadhesion agents, they should not be applied to molds for parts that are to be painted or plated, and the spray should not be used in plants where painting operations are performed unless steps are taken to prevent the silicone mist from entering the air-circulation system. Once in the air-circulation system of the plant, it may be carried long distances and may interfere with the painting operations.

Silicone release agents are also supplied in liquid form; they can be applied with a brush or cloth to a particular area of the mold. This method avoids the dispersion of silicone mist in the air. A formulation of silicone mold-release agent is available that does not interfere with painting; unfortunately, however, it is not nearly as effective or long lasting as the regular mold-release formula. The silicone mold-release agents can be used on a wide range of plastics but may cause appearance problems with clear polycarbonates and acrylics.

Zinc stearate powder is a very effective mold-release agent for the clear polycarbonates and acrylic resins as well as for many other plastics. It is usually applied from a tightly woven fabric bag: the bag can be struck against the mold, causing a fine coating of dust to cover it, or a stream of air from a compressed air source can be directed along the bag and at the mold. The particular disadvantage of zinc stearate, although it is an excellent release agent, is that it is easy to overapply it and build up a sticky deposit on die faces, machine strain rods, and molded parts at nearby machines. Zinc stearate can also be obtained in suspension in aerosol dispensing cans, but the nozzle orifice may clog occasionally.

Other mold-release agents that give good results are: (a) a dispersion of fluorocarbon that can be used with acrylic resins; (b) a neutral oil spray for use on polyethylene, nylon, polycarbonate, acetal, and poly(vinyl chloride); and (c) a vegetable oil-base spray for application in the food industry and where silicone interferes with end-product use. Because of the many variables, the best lubricant for a given purpose must be determined by actual trials. See also RELEASE AGENTS.

Purging. Changing from one material to another or from one color to another in an injection-molding machine can be costly and time consuming, depending upon the amount of contamination that can be tolerated. If the new material has a higher melting point, it may clean out the old material without trouble; several cycles of high and low cylinder temperatures usually help. Stubborn cases are sometimes helped by raising cylinder temperatures and purging with ground acrylic sheet scrap, which has more scouring action than most plastics. This may remove "hang-up" from previous materials, but the acrylic must then be purged with the new material. A number of commercial preparations for purging are available.

It is helpful to keep a log book on each machine, recording each change in color or material, problems encountered, and the final outcome. A pattern that changes noticeably may indicate internal cracks in the heating cylinder that will continue to give trouble. Since the pattern will make possible quick recognition of the problem, the cylinder can be cleaned or repaired immediately, thus avoiding loss of time and material.

Preventing Rust on the Mold. Opening the mold and ejecting the part results in a wiping action of the mold surfaces that leaves them vulnerable to tarnish and rust, particularly if mold lubricants are not used. This action is more rapid in humid weather, especially when cool water has been circulated through the die. Under these conditions, the circulating water should be turned off or even disconnected before molding is stopped, so that water cannot leak through an ill-fitting valve and chill the mold below the dew point. Failure to prevent rusting can result in a costly repolishing job and attendant loss of production, or even in complete loss of the cavity insert if the repolishing affects dimensions or optical properties.

Rust-prevention measures should be taken on any mold before storage or even short-term shutdown in the press. (A surprising amount of damage can be done overnight or during a weekend.) With the mold at approximately room temperature, it should be sprayed with a rust-preventive coating such as a wax in a petroleum solvent. Most oil companies have preparations suitable for this use.

Procedures Associated With the Molded Part

Finishing, Degating, and Trimming. The importance of the finishing, degating, and trimming operations should not be overlooked, nor should the proportion of their cost to that of the total cost of the molded part. For many years these operations, along with the packing in many cases, have been more or less ignored as cost items because they have been performed by the press operator while waiting for the press to produce the next cluster of parts.

With the increase in automation, which results in the operator being responsible for a number of presses, the operator may no longer be able to perform finishing operations. Also, automatic devices for degating the part while ejecting it from the mold may require inspection of every part if there are imperfections in the mold. These

Fig. 10. Automatic machine to deflash, drill, chamfer, and buff cord hole openings of telephone handle. Courtesy Western Electric Co., Inc.

factors may result in an increase in direct labor required at the press and thus in an actual increase in unit cost for the finishing operations.

The most useful tool for manual degating at the molding press is a pair of side cutters. These can be obtained with one side ground flat so the gate can be cut flush with the part. Cutters come equipped with adjustable stops to prevent damage to the cutting edges and with replaceable steel jaws or carbide-tipped jaws for glass fiber-filled and other abrasive materials.

A knife is normally used for trimming flash from molded parts. A commercial knife handle with a slot through it to accept a steel blade is a handy and inexpensive tool, since the blade can be ground to the particular shape and sharpness suited to the job. The blade can be moved forward as it wears, locked in place with a set screw, and quickly and inexpensively replaced when it wears out. Other hand finishing at the press is done with sandpaper fastened to a flat plate, files, drills, and chamfering tools

Fig. 11. Automatic machine for removing parting-line flash and restoring finish of telephone handle. Courtesy Western Electric Co., Inc.

set in handles. General-use saws and drills may be used on plastic; however, for high-production jobs these tools should be specially ground and operated at speeds and feeds suited to the material. See also MACHINING.

Operations such as broaching holes with an arbor press or an air cylinder, drilling or reaming with a drill press, and polishing with a motor-driven polishing wheel can also be done at the molding press. If the appearance of the part is important and it can be packed at the press, it may be economical to station a helper at the press to complete these operations during the press cycle.

In addition to the facilities mentioned, a well-equipped finishing department will have speed lathes for removing flash or ring gates from circular parts and tapping machines for tapping holes. If precision parts are to be produced from acetal resins, a controlled-temperature annealing bath usable to about 325°F should be included. Tumbling barrels are frequently used to remove flash and to polish the pieces. If finish is important, a large wooden barrel loaded with a large volume of wooden pegs and fine abrasive creams is used. Different plastics and abrasives behave so differently, however, that it is best to consult the supplier of both the plastic and the abrasive for recommendations.

Finishing can be automated when large numbers of parts are involved and when uniformity of finish is important. This is particularly true if the mold parting lines are difficult to maintain. Such a part is the telephone handset handle that has a flash where the two cores meet in the middle of the handle and a parting-line flash that is an irregular curve. In addition, the part is gated where the cord opening is to be located. The finishing is done on two automatic machines, each of which is loaded by an operator. The machine, shown in Figure 10, has an indexing table with fixtures that grip the transmitter end of the handle. One position has a curved rod that is pushed inside the handle on an arc to cut and remove the flash so that wires can be pushed through the handle when the handset is assembled. At subsequent positions the cord hole is drilled, chamfered, and buffed; the plastic shavings are then vacuum exhausted and the fixture opens for removal of the handle. The operator places these handles on a conveyor that takes them to an automatic buffing and polishing machine, shown in Figure 11. Here, the handles are placed on fixtures that ride around a long track, presenting the handle to various buffing and polishing wheels where any flash is removed and the finish restored. Although machines of this type are costly, they save investment in the additional molding facilities that would be required if there were more downtime for repairing molding tools. Also, tool-repair effort is less since the molds need not be maintained in as good condition if the handles are to be machine finished.

Another method of finishing is to bombard the plastic parts with small lightweight particles at very high speeds. Ground corn cobs and crushed apricot pits have been widely used since they have little effect on the flat surfaces of parts. Recently, small particles of polycarbonate resin approximately 0.050 in. in diameter by 0.050 in. long have been used with success. This method is inexpensive and is widely used on parts made of phenolic resins or other relatively brittle plastics; however, it is not of much value on the tougher thermoplastics. Small parts that can be tumbled gently

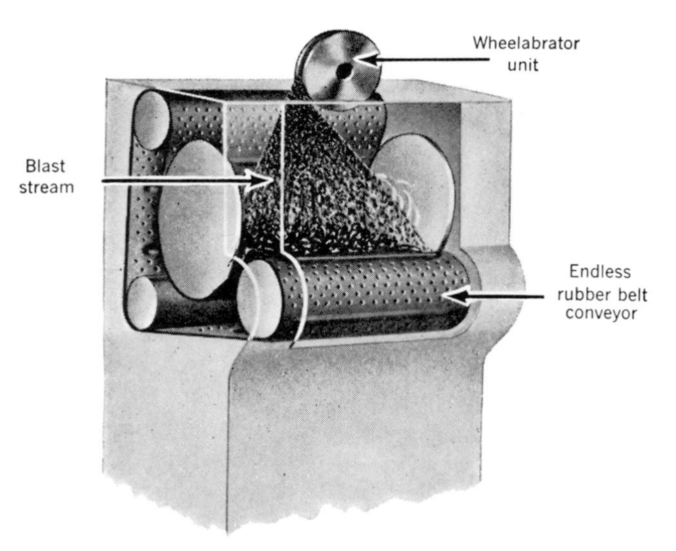

Fig. 12. Batch-load deflashers, designed to handle large volumes of parts not too thin or fragile to withstand gentle tumbling are available in 2- or 5-ft³ operating-load capacities. Blast cycles are generally less than 6 minutes, and variable controls are available. Courtesy Wheelabrator Corp.

are usually cleaned in a machine such as that shown in Figure 12, where the parts are rolled over on an endless belt while bombarded from above. Larger parts that might be damaged by the tumbling action can be placed on rotating tables, as shown in

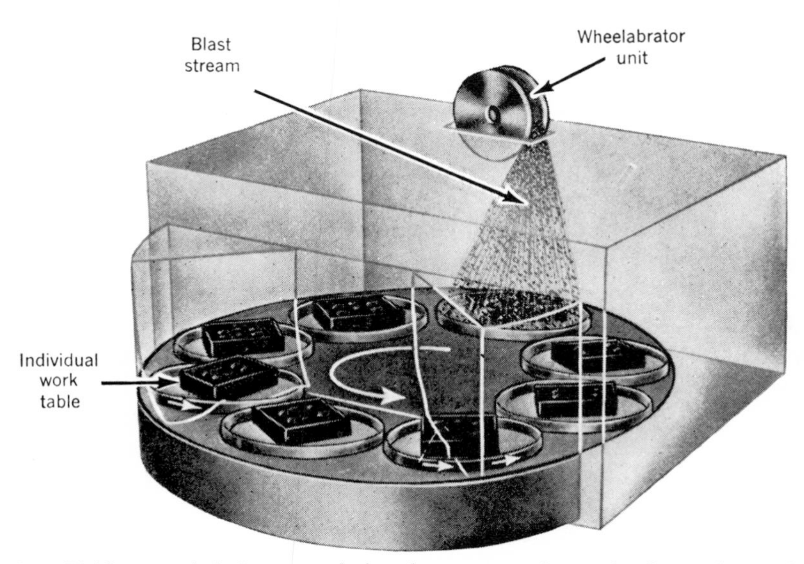

Fig. 13. Table-type deflashers are designed to process large, fragile, and complex molded shapes. Tables are equipped with various combinations of independent rubber-covered work tables which rotate within the blast zone. Courtesy Wheelabrator Corp.

Figure 13, and bombarded. It may be necessary to turn the parts manually to do a thorough job.

No finishing department is complete without pedestal buffers and bench buffers. Conditions that create surface finishes which require buffing of the part are: (a) inadequate polish on mold cavity, (b) tarnished or damaged area in mold cavity, (c) blush on a piece part due to molding conditions or inadequate venting, (d) blushed area in the mold caused by conditions in (c), and (e) damage to parts caused by accidents or careless handling. Whether the condition affects a small percentage of the product or the entire batch, there will be occasions when a little buffing will save many times its cost in the value of molded parts.

Buffing and polishing are frequently done on a ventilated wheel of open construction made up of alternating layers of 12-in. and 6-in.-diameter muslin discs. Depending upon the plastic and the application, a suitable buffing compound can be selected from the large number of commercial compounds available.

Decoration. The growth of decorative applications of plastic parts has been as impressive as that of the plastics industry itself. Useful as well as tasteful examples of decorated parts are commonplace. Some of the decorating methods in common use are: hot stamping, wiping paint into molded letters on designs, molded-in decals, painting with organic coatings, chrome plating and other metallic plating, vacuum metallizing, silk screening, dip coating, and roller coating.

Because of the many problems created by the variety of plastic materials, solvents, pigments, surface preparation, surface contaminants, and mold lubricants, information on specific applications should be obtained from materials suppliers. See also Coating methods; Decorating; Metallizing.

Assembly Operations. The finishing department is often the logical place to perform assembly operations that are peculiar to plastics, such as solvent bonding, spin welding, and ultrasonic welding. Solvent bonding with suitable agents and techniques develops bond strength that approaches the strength of the material being bonded. The selection of solvent depends upon the plastic. Literature furnished by materials suppliers usually lists the solvents recommended, percent of plastic to be dissolved, and other pertinent information. Materials such as the acetal resins have such excellent solvent resistance that it is almost impossible to use solvent bonding (see also ADHESION AND BONDING).

If the parts are of such a shape that one can be rotated in contact with the other, frictional heat is developed at the interface and they can be spin welded. This method is fast, simple, and develops excellent welds. Drill presses and speed lathes can usually be adapted for spin welding.

Ultrasonic welding is another method of joining rigid structures of various plastics that has been developed in recent years. It is coming into increasingly wide use because it requires no messy cements and toxic solvents and it can be used regardless of the shape of the parts. Not only can two pieces be bonded together, but metallic inserts can be inserted into a plastic part after molding. The actual welding is often done in a fraction of a second, but obviously the time for the total operation will depend on the assembly fixtures used.

Reclamation and Reuse of Material

The average injection-molding or blow-molding operation produces considerable amounts of purgings, sprues, runners, pinch-offs, and rejects, and it is important to find ways of reusing this material. Because a relatively high portion of the total cost of a part is the material, a seemingly small amount of lost material can appreciably increase the cost of the part. Most thermoplastics can be reused several times, assuming they have not been "abused." The material supplier should be consulted regarding any critical properties.

Before the material is put into a reusable physical shape, the following factors should be considered:

1. Purgings obtained when changing from one material to another are rarely usable, but purgings produced by changing from one color to another in the same grade of material may be usable when color is not important. Purgings from overheated cylinders or cylinders that have stood too long at the molding temperature are likely to be degraded. Purgings resulting from the cleaning of a chamber that is producing streaks or specks in a clear material should not be used again except in parts where these are not objectionable.

2. Rejects must be treated carefully. The material in a part rejected because of dirty streaks or specks should not be reused, as it might contaminate larger quantities of material. The material in parts rejected for dimensional defects or for sink marks or underfills can safely be reused.

3. Sprues and runners are generally usable when the press is running normally. Sprues and runners from clear parts such as the acrylic resins, where appearance is usually critical, are frequently fed back through a port at the rear of the heating cylinder to minimize the chances of contamination.

4. Pinched-off material from blow-molding operations is usually granulated and fed back to the machine.

There are a number of methods of preparing these materials for reuse, and the method selected depends upon the economic factors of the individual operation, such as the amount of material to be handled, the cost of facilities, the amount of labor involved, and the effect on the quality of the material.

As previously mentioned, the simplest way of reclaiming material for the plunger-type press is to feed the sprue and runner back through a port at the rear of the heating chamber. It may be necessary to cut the runner into several pieces to feed it through the port. This assumes that an operator is attending the press and that he has time for the reclaiming task. It is important that this task be performed while the sprue is still warm and that a uniform material feed be maintained. This method cannot be used on a reciprocating screw press which requires granular material. In either case it may be economical to provide a small granulator at the press to handle not only sprues and runners but reject parts.

If the material is fed manually to the press, the material handler can mix the regrind with the virgin material before putting it in the hopper. An example of this type of granulator, shown in Figure 14, has a 5-hp motor and a rating of 150 lb/hr. Other types are available at various power ratings.

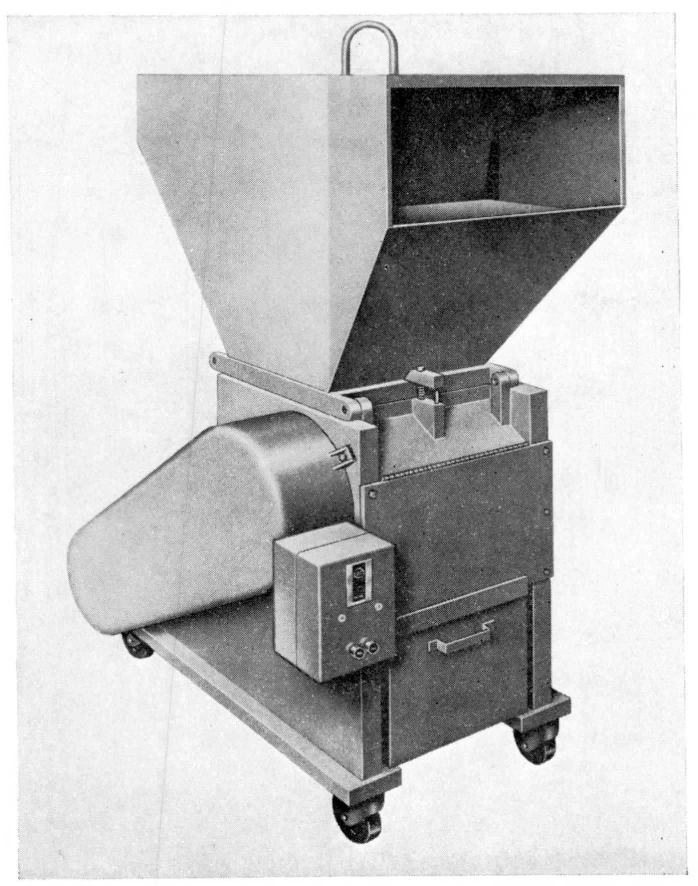

Fig. 14. Portable granulator with 10 × 14 in. throat opening for use beside the press. This granulator can be equipped with motors from 3 to 7½ hp. Courtesy Foremost Machine Builders, Inc.

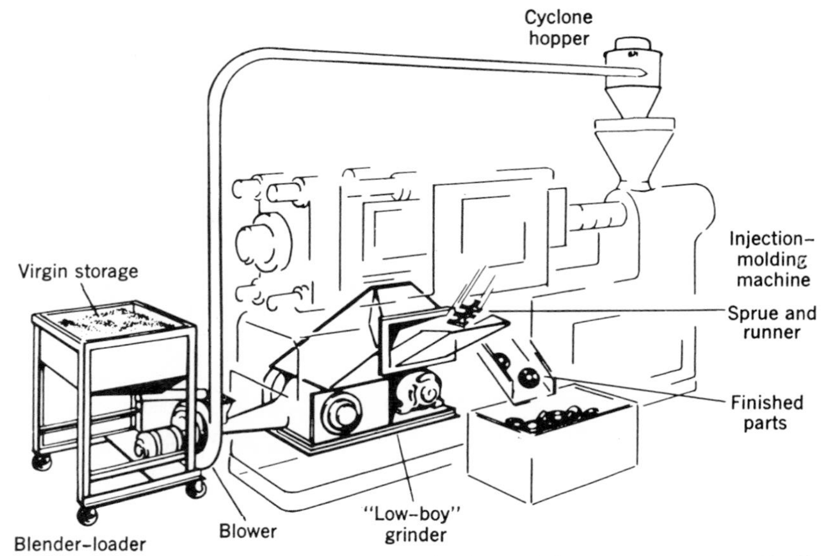

Fig. 15. Sprue and runner from automatic injection-molding press are automatically ground, mixed with virgin material, and fed back to the machine hopper. Courtesy Foremost Machinery Corp.

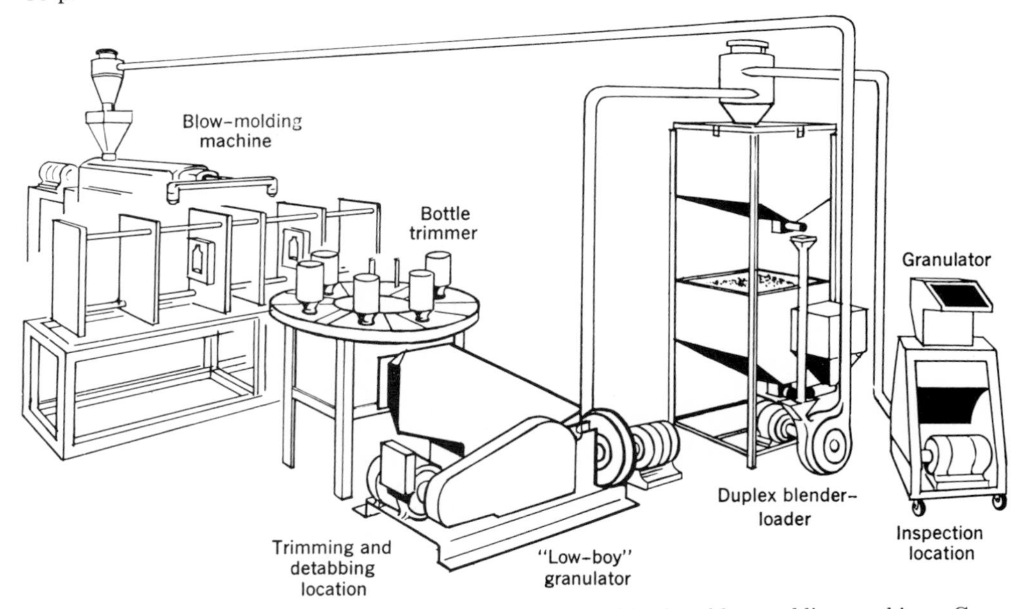

Fig. 16. Drawing showing how tabs and scrap bottles are fed back to blow-molding machine. Courtesy Foremost Machinery Corp.

Quieter granulators have been developed and are recommended for use where noise is a problem in the molding room. If the press is equipped with a pneumatic or vacuum loader, a conveyor can be connected directly to the granulator bin and the regrind conveyed back to the hopper for reuse. This method tends to proportion the regrind so that the ratio of virgin material to regrind is held fairly constant.

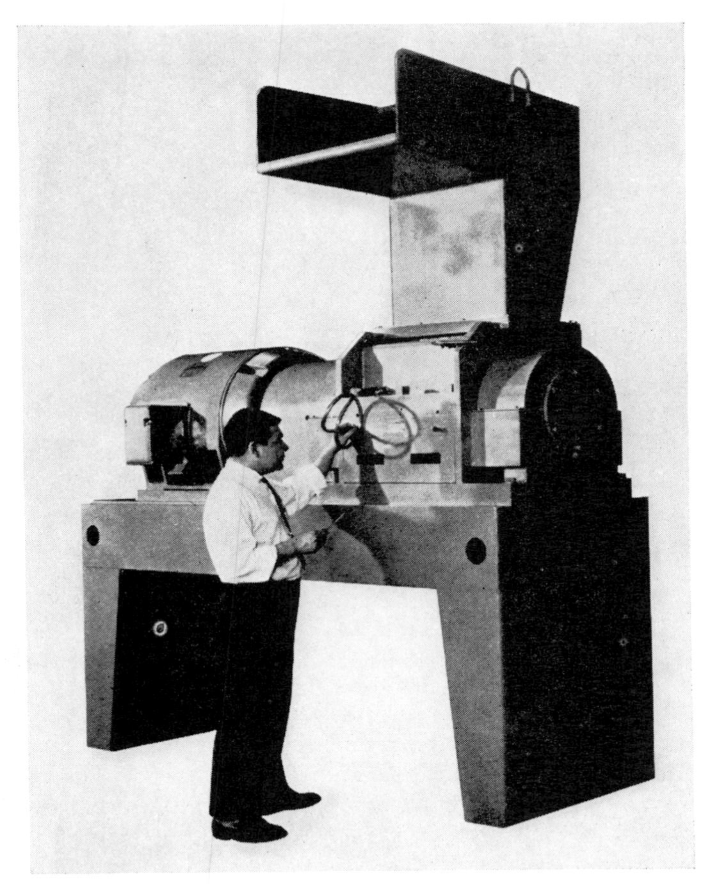

Fig. 17. A large, heavy-duty granulator for centralized material-reclamation system that can also be used for granulating purgings. Horsepower ratings are up to 250. Courtesy Entoleter, Inc.

As more presses and tools are designed for full automatic operation, it becomes possible more often to fully automate the job, including material handling and scrap reclamation. Whenever the machine and tool can be designed to degate the parts in the mold and eject the parts separately, the sprue and runner system can be ejected into a chute and carried by gravity to a "low-boy" granulator, from which the material is conveyed back to the machine hopper, as shown in Figure 15.

For blow-molding applications, the reusable material may consist of trim from the neck of the bottle and a tab from the bottom. The usual setup is such that the trimmings fall by gravity into a "low-boy" granulator located under the trimming machine. Rejected large bottles require a different design of granulator which may be located at the inspection or sorting position. The regrind from both locations is fed to a common hopper. Because it is important that the ratio of virgin material to regrind be held constant in blow molding, a duplex blender–loader is employed. A schematic drawing of this setup is shown in Figure 16.

Centralized material reclamation systems are frequently employed where a number of presses are using the same material in relatively large quantities. Instead of providing a granulator at each press, the material can be accumulated in a central location

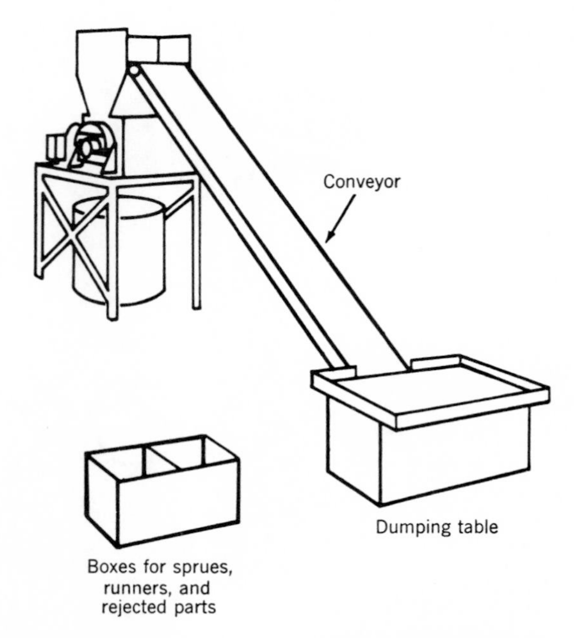

Fig. 18. Dumping table and conveyor for use in centralized material-reclamation system. Courtesy Entoleter, Inc.

and granulated in a large heavy-duty granulator such as the one shown in Figure 17. Granulators of this size have motors ranging from 30 to 250 hp, depending upon the material to be granulated and its form, and may have outputs as high as 2500 lb/hr. A dumping table and conveyor, such as that shown in Figure 18, are often used to feed the granulator if the material to be granulated tends to be bulky. The conveyor on an installation like this should be interlocked with the granulator motor to stop the conveyor when the granulator motor approaches full load. This type of granulating setup is difficult to clean; its use is not advisable where frequent color or material changes are necessary. Material from this granulator may be either reused as regrind or mixed with virgin material in a desired proportion and reused, depending upon the requirements of the application.

Because granulating generally produces irregularly shaped particles and more fines than are present in virgin material, it may prove expedient to repelletize this material for critical applications. This is not as likely to be required for a reciprocating-screw injection-molding machine as for a plunger machine. The usual means of pelletizing are to extrude the granulated material in ⅛-in.-diameter strands and to chop them to ⅛-in. lengths in a pelletizer. Many commercial extruders are available; selection should be based on the material to be extruded and the quantities involved.

As in the case of the large central granulator, cleaning the extruder is a time-consuming process when materials or even colors in the same material are changed. If more than one material is to be reclaimed in this way, it is necessary to accumulate each material and run it through separately. It is also essential to prevent other materials or colors from getting mixed in accidentally, because a small amount of contamination may spoil a large lot of material. Many materials change color slightly, depending on the time and temperature associated with the molding operation. If color matching of various parts of an assembly is critical, parts molded from regrind may have a slightly different color because of the second heating, and parts molded from extruded regrind may be even more different in color because of the third heating. This effect varies widely with the material, the color, and the equipment used, and several trial runs should be made before deciding on the method to use.

Purgings can be granulated in many of the larger granulators, particularly if care is taken to see that purgings are not permitted to become unnecessarily large and if they are loaded judiciously so as not to overload the granulator. It is important in all granulators that the blades be kept sharp and set with the proper clearance if a good job of granulating is to be done with a minimum of fines and heat buildup and without overloading the motor. If there is no need for a large granulator other than to grind purgings, it would be wise to consider a two-stage granulator. The first stage is a rotating drum with teeth, which reduces the purging to chunks that can be granulated by the second stage (a smaller conventional granulator), thus accomplishing the purpose with a lighter machine and less horsepower.

Bibliography

Buyer's Guide to Silicone Spray, Injection Molder's Supply Co., Cleveland, Ohio.

Painting Plastics Manual, Bee Chemical Co., Lansing, Ill., 1962.

Standard Injection Machine Procedures, Injection Molder's Supply Co., Cleveland, Ohio.

J. Boyden, "Auto Scrap Recycling," in *Society of Plastics Engineers Technical Papers, 20th Annual Technical Conference, January 27–30, 1964,* Society of Plastics Engineers, Stamford, Conn., Vol. X, Section X-4, pp. 1–4.

R. H. Dean, "Temperature Control as Applied to the Plastics Industry," in *Society of Plastics Engineers Technical Papers, 18th Annual Technical Conference, January 30–February 2, 1962,* Society of Plastics Engineers, Stamford, Conn., Vol. VIII, Section 16-2, pp. 1–6.

S. E. Giragosian and G. G. Freygang, "Some Causes and Cures of Splash Defects in Acrylic Molding," in *Society of Plastics Engineers Technical Papers, 16th Annual Technical Conference, January 1960,* Society of Plastics Engineers, Stamford, Conn., Vol. VI, Section 13-II, pp. 1–3.

W. H. Meyer, "Mold Cooling and Temperature Control for Polypropylene," in *Society of Plastics Engineers Technical Papers, 19th Annual Technical Conference, February 26–March 1, 1963,* Society of Plastics Engineers, Stamford, Conn., Vol. IX, Section VI-3, pp. 1–10.

H. A. Meyrick, "What You Should Know About Mold Cooling," *Mod. Plastics* **41,** 219 (Oct. 1963).

L. Temesvary, "Mold Cooling: Key to Fast Molding," *Mod. Plastics* **44,** 125 (Dec. 1966).

C. E. Waters, "Water Conservation and Chilling as Applied to Plastic Processing," in *Society of Plastics Engineers Technical Papers, 18th Annual Technical Conference, January 30–February 2, 1962,* Society of Plastics Engineers, Stamford, Conn., Vol. VIII, Section 16-3, pp. 1–5.

D. C. Whitlock, "Automatic Conveying of Plastic Materials," in *Society of Plastics Engineers Technical Papers, 22nd Annual Technical Conference, March 7–10, 1966,* Society of Plastics Engineers, Stamford, Conn., Vol. XII, Section XI-4, pp. 1–9.

H. Whitlock, "In-Plant Material Handling and Drying," in *Society of Plastics Engineers Technical Papers, 18th Annual Technical Conference, January 30–February 2, 1962,* Society of Plastics Engineers, Stamford, Conn., Vol. VIII, Section 16-1, pp. 1–2.

H. Y. Yates, "Bulk Storage," in *Society of Plastics Engineers Technical Papers, 20th Annual Technical Conference, January 27–30, 1964,* Society of Plastics Engineers, Stamford, Conn., Vol. X, Section X-1, pp. 1–4.

A. B. Hitchcock
Western Electric Company, Inc.

BAG MOLDING

Bag molding is a very versatile process of manufacturing reinforced plastic parts. It is a procedure in which a combination of a reinforcement and a thermosetting resin is placed in a mold and covered with a flexible diaphragm. Heat and pressure are then applied to cure the reinforced plastic materials to a desired configuration. Although the bag-molding technique has been in use more than twenty years, it is still considered "art." There are many companies that have well-trained production personnel and an aggressive reinforced-plastic development program. As a result, bag-molding techniques are continually being improved, and new high-strength, high-temperature, reinforced plastic materials are being successfully bag-molded for the aircraft and missile industries. One of the latest advancements has been the fabrication of large high-temperature, sandwich-type nose radomes ten feet in diameter and over twenty-five feet long. The radomes consist of high-temperature, glass-reinforced plastic skins with a high-temperature, glass-reinforced plastic core. This process is recommended primarily for prototype parts, parts with small production runs, large or complex shapes, and parts that require high strength and reliability. The size of a part that can be made by the bag-molding process is limited only by the size of the mold and size of the curing oven or autoclave.

General Description of the Process

The general process of bag molding can be subdivided into three specific molding methods: vacuum bag, pressure bag, and autoclave (see Fig. 1).

The vacuum-bag and autoclave methods are the most popular and are the methods used to produce the vast majority of bag-molded parts. The advantages of the vacuum-bag and autoclave methods are that the tooling is relatively inexpensive and that the basic equipment (oven and autoclave) can be used for an unlimited variety of parts. The main disadvantage of the pressure-bag system is that the tooling and equipment are integral and relatively expensive because they can only be used for the part for which they were designed. These parts are economical only for large production runs. Therefore, the emphasis of this article will be placed on the vacuum-bag and autoclave methods. The same principles can be applied to the pressure-bag method with minor variations.

Process Steps. The general steps in the bag-molding process are as follows:

1. The reinforced plastic materials (usually some form of glass reinforcements and a thermosetting resin) are laid up in a male or female mold.

2. The materials are then covered with a flexible film diaphragm (usually called a vacuum bag) and the film is sealed to the mold.

3. Pressure is applied to the laid-up materials by creating a vacuum between them and the flexible diaphragm. If additional pressure is desired (as in the case of auto-

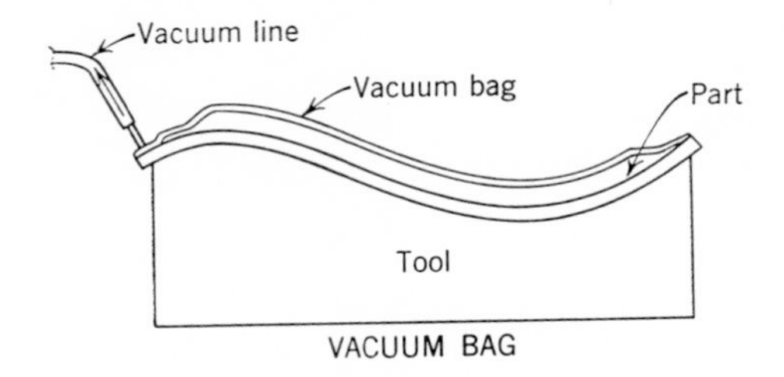

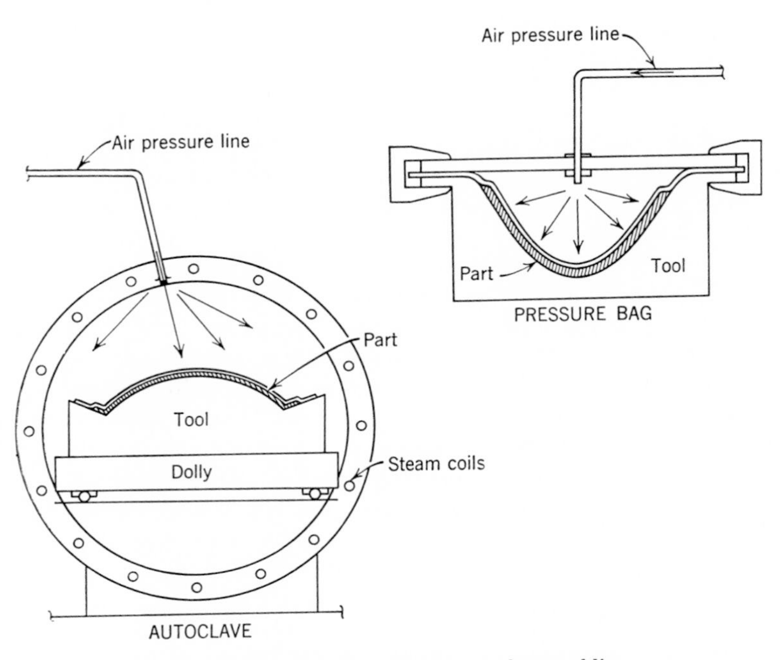

Fig. 1. Specific molding methods in bag molding.

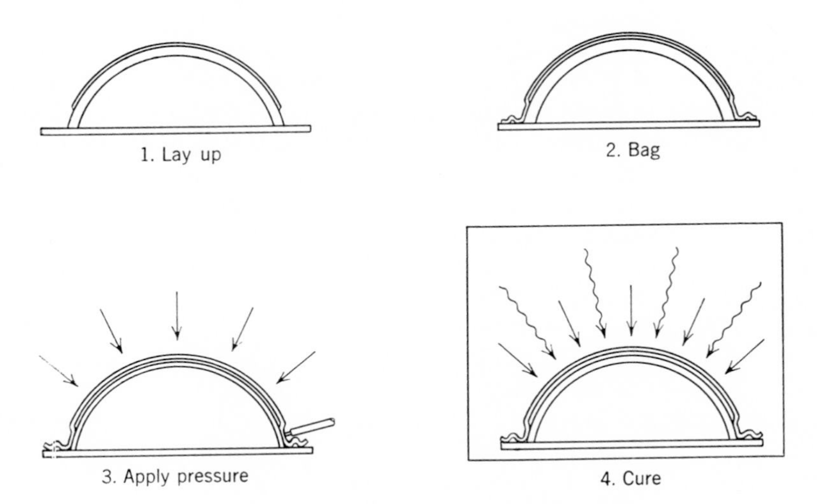

Fig. 2. General process steps in bag molding.

clave molding), external air pressure is applied to the flexible diaphragm, which in turn transmits the pressure to the laid-up material beneath.

4. The reinforced materials are then cured under pressure by the application of heat and the chemical action of the catalyst or curing agent in the resin (see Fig. 2).

Materials Used in Bag Molding

Primary Materials. Thermosetting resins and fibrous reinforcements are the primary materials of reinforced plastic parts. The reinforcement supplies the struc-

Table 1. Resins Used in Bag Molding

Resin type	Military specification	Operating temp, °C[b]	Approx cost: As liquid resin; $/lb	Approx cost: As prepreg glass fabric[a]; $/yd²	Typical uses
general-purpose polyester	MIL-R-7575	82	0.43	1.80	boats, luggage, radomes, prototype parts
modified polyester	MIL-R-25042	260	3.50		high-temperature radomes
epoxy	MIL-R-9300	260	0.65	2.50	radomes, aircraft components, chemical tanks
phenolic	MIL-P-25515	260		2.00	high-temperature aircraft and missile components
silicone	MIL-P-25518	260		3.20	extremely high-temperature applications

[a] Prepreg: Preimpregnated glass fabric, style 181, in production quantities.
[b] Limit for continuous exposure. Materials can be used at higher temperatures for short time.

Table 2. Reinforcements Used in Bag Molding[a]

Type	Military specification	Typical form supplied	Typical styles	Cost, $	Typical uses
glass mat chemical binder mechanical binder	MIL-M-15617	made from chopped strands and contínous strands; supplied in roll up to 76 in. wide, 200 ft long		0.60/lb[b]	used primarily with general-purpose polyester resin where cost is important and medium structural properties are required
glass fabrics	MIL-F-9084	made from continuous filament yarns, rovings, and staple yarns; standard size roll: 38 in. wide, 200 ft long	120 143 181 182 183	0.70/yd² 1.10 1.18 1.50 1.88	used with all types of resin systems when high structural properties are required

[a] Other types of reinforcements that are occasionally used are asbestos felts, high-silica glass, quartz, and graphite fabric. Glass and aluminum honeycomb cores are used in conjunction with reinforced-plastic faces for radomes or lightweight sandwich parts.
[b] Approximate figure.

tural characteristics while resin binds the reinforcements together in the desired configuration. The most common materials used are shown in Tables 1 and 2.

Secondary Materials. Secondary materials are those which are used in the manufacturing of reinforced plastic parts, but do not become a part of the finished part. These materials are directly related to the steps in the bag-molding processes. The success of the molding operation is determined to a great extent by how well these materials perform their specific functions. The most common secondary mate-

Table 3. Secondary Materials

Basic materials	Types	Approx cost, $	Functions
mold-parting agents	silicone liquid and pastes[a]	4.80/lb	to ensure release of part from mold
	high-temp waxes and greases	4.00/gal	
	soluble films in solution	3.50/gal	
films	poly(vinyl alcohol)	1.70/lb	to ensure release of bleeder material from part, and to form a flexible diaphragm (bag) over lay up during cure
parting films	poly(vinyl fluoride)	4.70/lb	
bag films (flexible diaphragm)	cellophane	0.78/lb	
	polyester (Mylar)	2.35/lb	
	poly(vinyl chloride)	0.66/lb	
	polyurethan	2.40/lb	
	silicon rubber sheet	7.00/lb	
bleeder materials	glass cloth	0.75/yd	to provide pathway for getting air and excess resin out of part
	cotton cloth	0.35/yd	
	glass mat	0.50/lb	
	jute	0.20/lb	
bag-sealing compound	zinc chromate plastic sealing compounds	1.00/lb	to seal flexible diaphragm to mold; provides partial void area between flexible diaphragm so atmospheric pressure is applied to part during cure

[a] Do not use if part is to be painted after molding.

rials and their functions are shown in Table 3. Figure 3 illustrates the general application of these materials. The following information can be used as a guide in selecting the proper secondary materials.

1. The selection of the best parting agent must take into account the type of mold material, the type of resin being used, the curing temperature of the part, and the surface treatment desired for the finished part. In some cases a combination of parting agents is used for best results.

2. The selection of the parting film and bag film must take into account the type of resin, the curing temperature and pressure, the contour of the part, and the shrinkage characteristics of the various types of available films.

3. The upper curing temperature of poly(vinyl alcohol), polyester, and cellophane films is 350°F and of poly(vinyl chloride) is 300°F. Poly(vinyl alcohol) film has the greatest shrinkage during curing, whereas poly(vinyl chloride) stretches. Cellophane and polyester have a small amount of shrinkage. The selection of the overall bleeder material should take into account the desired surface condition of the bag side of the finished part. The pattern of the bleeder material will be imprinted on the bag-surface side of the part.

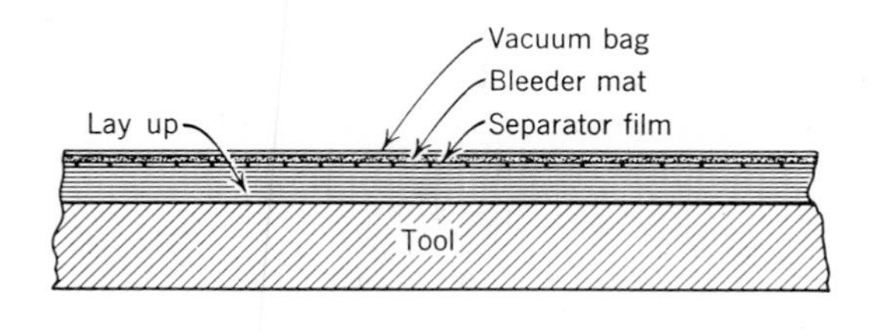

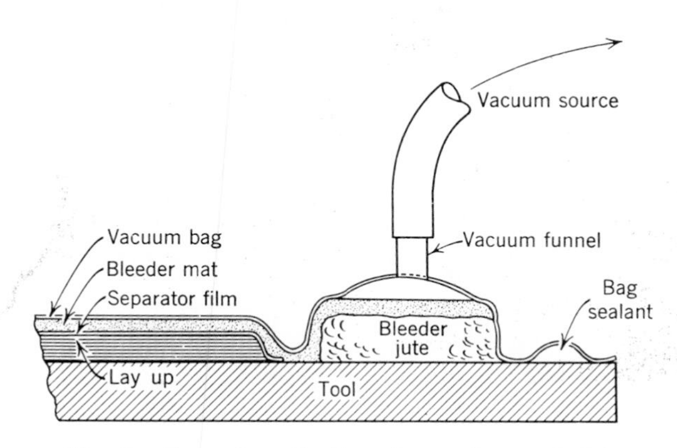

Fig. 3. General application of secondary materials.

Basic Equipment

Adequate equipment and proper maintenance is essential in maintaining the production of high-quality, bag-molded, reinforced plastic parts. Early in the development of bag molding, the selling point of this process was that it required very little basic equipment. As competition arose, the bag-molding processes were improved, and more emphasis was placed on equipment. It soon became evident that adequate equipment was absolutely necessary to meet competition and to stay in business. The essential basic equipment required for an up-to-date plastic shop is listed below.

Vacuum System. The vacuum system is the heart of the bag-molding operation. There should be a minimum of two pumps. Each pump should be of sufficient capacity to maintain a vacuum-line reading of 21–28 in. of mercury under normal operating conditions. The second pump should be connected so that it will automatically go into operation when the line reading gets below 21 in. of mercury. A number of vacuum gages should be strategically located throughout the fabricating and curing areas so that the vacuum pressure is clearly visible in all areas. Their accuracy should be checked at least every three months.

Oven. The oven is the center of the vacuum-bag curing system. It may be heated by electricity or gas. A forced-draft circulating system is necessary to insure a uniform distribution of heat. The temperature capability range should be from room temperature to 500°F with an accuracy of ±15°F. The temperature should be automatically controlled and recorded. Typical production-oven sizes are 8 × 12 × 12 ft, 8 × 12 × 30 ft, and 10 × 12 × 24 ft. Each oven should have its own vacuum manifold system, and each individual vacuum line should have its own vacuum gage located on the outside of the oven (see Fig. 4). The system should be designed so that if a

Fig. 4. Vacuum curing oven. Courtesy General Dynamics Corp.

part in the oven loses its vacuum. it can immediately be detected by a drop in its respective vacuum gage.

Infrared Heater. A bank of infrared lights (connected to a timing mechanism for accurate temperature control) may be used instead of an oven for curing bag-molded parts. It is essential to keep the part from coming too close to the heat sources to avoid local overheating.

Autoclave. The autoclave is the center of the curing system when more than atmospheric pressure is required on a bag-molded part. The autoclave may be heated by electricity or steam coils inside the steel pressure chamber. A large fan, or another method of air circulation, is used inside the steel pressure chamber to ensure an even distribution of heat. The advantage of using steam is that the autoclave can be cooled quickly by running water through the steam coils. Temperature and vacuum controls similar to those used on curing ovens are required. Additional pressure controls are also required. Because of the high structural requirement of the autoclave in comparison to a typical heating oven, the cost per cubic foot of working space of an autoclave is many times higher than the cost of a good-quality heating oven.

Compressed-Air System. A typical production-type compressed-air system will supply all the air required for the average bag-molding production shop. If a large autoclave is used it may be necessary to have a separate air compressor just for the autoclave. The compressed-air system should preferably include a reserve air compressor that will automatically be connected to the system, so that a line pressure between 120 and 150 psig is maintained.

Trimming and Drilling Equipment. For ease of lay up, most bag-molded parts are made with an excess of material beyond the net size of the part. Therefore, it is very important to have adequate facilities and equipment for the trimming and drilling operation. A completely enclosed room should be provided to prevent the dust from contaminating the other areas. Each machine should be connected to a dust-collecting system because the dust can cause local irritation to the eyes and skin. Because of the highly abrasive characteristics of glass-reinforced plastic, special cutting materials are used. Diamond cutting edges on circular saw blades, routers, and other equipment, have proved to be economical, even though the initial cost is high. Other cutting materials that are used are carborundum, carbide, and high-carbon steel. A typical list of trimming and drilling equipment is shown below:

1. Table saw; 10 in. blade, 48 × 48 in. tabletop.
2. Band saws; two machines with 12 in. throat and one machine with 36 in. throat.
3. Router; three machines, 36 × 36 in. tabletop.
4. Sanding machines; two machines of circular type with 24 in. diameter, two machines of belt type, 3 in. wide.
5. Miscellaneous hand tools (air-driven), grinders, sanders, routers, drill motors.

Tooling

One of the factors that contributes to the versatility of the bag-molding process is the large variety of tooling methods that can be economically used. When selecting the type of tooling, the following factors should be taken into consideration: the number of parts to be produced, which surface of the part is most critical, the type of resin system to be used, and the method of molding.

Basically, bag-molding tooling is limited to one surface of the part, either male or female. The other surface will be formed by the flexible film diaphragm. Temporary tools are usually made from wood or plaster. Very few parts can be produced from these temporary molds if the parts require curing in an oven; more parts can be produced on this type of tooling if the parts can be cured at room temperature. Permanent tooling is made of reinforced plastics, aluminum, or steel. The most versatile and economical mold material for moderate production runs (up to approximately 400 parts) is epoxy-reinforced plastic. These tools are usually made from plaster splashes taken from the tooling master form, which is the exact size and shape of the reinforced plastic part to be produced. The method of fabrication of the plastic tool is essentially the same as making a reinforced plastic part, except that it is approximately $\frac{3}{8}$–$\frac{1}{2}$ in. thick and has a mold-in base. Sheet aluminum ($\frac{1}{4}$–$\frac{3}{4}$ in.) is used to make molds having a gentle contour that can be formed by a rolling operation. If a mold is made of several pieces of sheet aluminum, great care must be taken to ensure that the welded joints do not leak. Some molds are made from aluminum castings, but it is difficult to obtain complex aluminum cast molds that have porosity-free surfaces. If large molds are to be made from aluminum, the difference in thermal expansion between aluminum and the reinforced plastic must be taken into consideration. Steel is used for molds (especially large molds) when extremely close tolerances are required, because the thermal expansion of steel is very close to that of reinforced plastics.

Assembly-Line Production in the Bag-Molding Process

The modern reinforced-plastic shop or department is organized around an assembly-line system. A definite area is assigned to each basic operation. The areas are located so that there is an efficient flow from the raw-material area to the final-inspection area of the completed reinforced plastic parts (see Fig. 5).

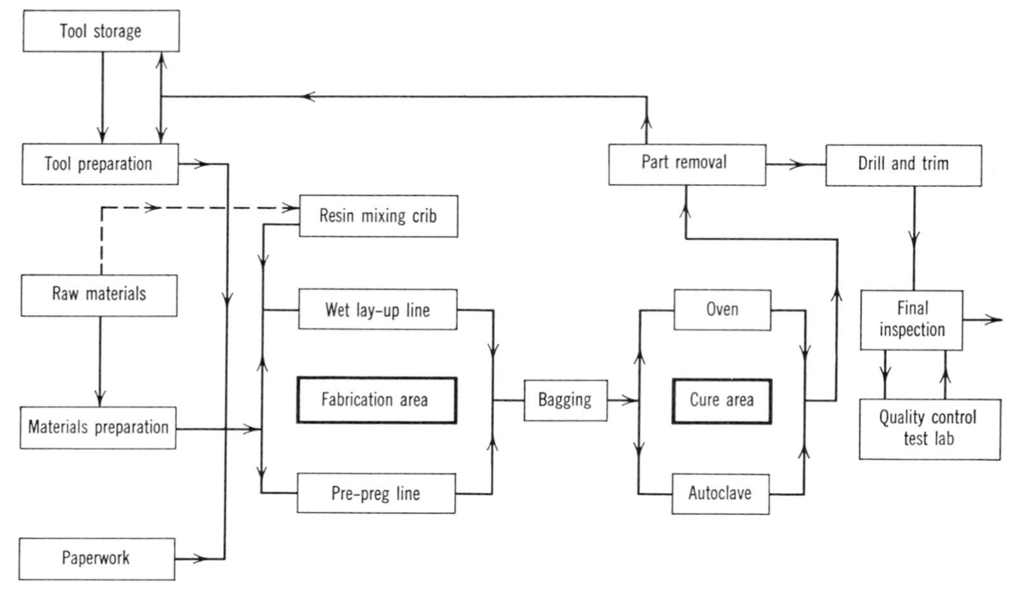

Fig. 5. Assembly-line production method in reinforced plastics bag molding. Area 1: tool storage, raw materials, and paperwork. Area 2: materials and tool preparation. Area 3: fabrication. Area 4: bagging. Area 5: curing oven and autoclave. Area 6: part removal, trim, and drill. Area 7: final inspection.

Basic Function of Each Assembly-Line Area

Area 1. Tool Storage, Raw Materials, and Paper Work. In this area tools are stored and necessary repairs and changes are made; glass cloth, glass mat, and resin are stored at room temperature; preimpregnated materials are stored under refrigeration; production paper work originates; and flat, rectangular patterns are developed for each piece of reinforcement, based on the engineering blueprint requirement (these patterns are called *cutting specifications*).

Area 2. Materials and Tooling Preparation. In this area tools are cleaned and inspected, parting agents are applied to the mold, and glass fabric, glass mat, or preimpregnated materials are cut for each part, as required by the applicable cutting specification.

Area 3. Fabrication Area. The tools, corresponding raw materials, and paper work are assembled together at the beginning of the fabrication area. If a part is to be made from preimpregnated material, it goes down the "preimpregnated" assembly line (see Fig. 6). If a part is to be made from dry-glass reinforcement and liquid resin, it goes down the "wet lay-up" assembly line. Before the dry glass is laid up, the correct amount of catalyzed resin is obtained from the resin-mixing crib. (Starting resin content for the mat is about 60% by weight, and for glass fabric 45% by weight.)

Basic lay-up steps using preimpregnated materials are as follows: (a) The first ply of preimpregnated fabric is placed (parting-film side up) on the mold by laying one end of the fabric down and working it smoothly toward the other end to avoid wrinkles. (b) The parting film is removed from the preimpregnated fabric and the material is worked against the mold with a Teflon or steel rubbing tool. A heat gun may be used to make the preimpregnated material more pliable, so it can be worked into sharp

Fig. 6. Fabrication assembly line for preimpregnated materials. Courtesy General Dynamics Corp.

radii. (c) The fabric is cut and darted when necessary in order to form the required contour. Overlaps of fabric are held to $\frac{3}{4}$ in. ± $\frac{1}{4}$ in. Overlaps should be kept to a minimum and never superimposed. (d) Steps a, b, and c are repeated until all plies of fabric have been laid up. (e) The part is now ready for the bagging operation.

Basic lay-up steps using the wet lay-up method (dry cloth and liquid resin) are as follows: (a) The prepared mold is coated with the catalyzed resin. (b) The first ply of fabric is placed on the mold by laying one end of the fabric down first and working it smoothly toward the other end to remove air and wrinkles. The fabric is worked until it adheres to the mold without air pockets, voids, or wrinkles. (c) The fabric is then given a brush coat of catalyzed resin. (d) The fabric is cut and darted to conform to the mold contour. Overlaps of fabric are held to $\frac{3}{4}$ in. ± $\frac{1}{4}$ in. Overlaps should be held to a minimum and never superimposed. (e) Steps b, c, and d, are repeated until the required number of plies of fabric have been laid up. (f) The part is now ready for the bagging operation.

There are four basic steps in fabricating a reinforced plastic sandwich part (see Fig. 7): (a) The top and bottom skins are laid up using either the wet lay-up method or the preimpregnated lay-up method. (b) The skins are bagged and cured. (c) The core is bonded to the bottom skin. (d) The edge members and top skin are bonded to the core and bottom skin. If preimpregnated fabric is used, these four steps may be combined into one operation; ie, both bonding and curing are carried out in one setup.

Area 4. Bagging. The method and materials that are used to bag a laid-up part are based upon the following: lay-up method—the preimpregnated system or the wet lay-up system, type of resin used, and curing temperature.

Fig. 7. Fabrication assembly line for large sandwich-type parts. Courtesy General Dynamics Corp.

A typical bagging system for a preimpregnated laid-up part is as follows (see Fig. 8): The lay-up is covered with a perforated parting film. The parting film should have sufficient number and size of holes to ensure adequate bleeding of air and excess resin out of the part, so that the finished part will have the desired glass-to-resin ratio. The correct size and location of the holes can only be determined by trial and error. The parting film is covered with a ply of bleeder material, extending it about 3.5 in. beyond the lay up. The bag-sealing compound is placed around the periphery and about 4 in. beyond the edge of the lay-up. The lay-up, parting film, and bleeder material are then covered with a flexible film diaphragm, which is sealed to the mold with the sealing compound. The vacuum hose is connected to the mold and pressure is slowly decreased by about 5 in. of mercury (this is known in the trade as applying "vacuum pressure"). It is held for 5 min and the wrinkles and excess air are worked out of the lay-up, bleeder material, and flexible diaphragm. Full vacuum is applied for 5 min and then a vacuum-leakage test is run. The vacuum valve is turned off and the gage watched. The maximum acceptable vacuum leakage rate is 1 in. of mercury in 1 min. If the leakage rate is greater than the maximum acceptable rate, the bag is checked for leaks. When the vacuum-leakage check has been passed, the part is placed on a portable curing rack, still under vacuum. It is now ready for curing in the oven or autoclave (see Fig. 9).

A typical bagging system for a wet laid-up part that must be rubbed out to be "void-free" (free as possible of entrapped air) is as follows (see Fig. 10): (a) The sealing compound is placed on the mold about 4 in. beyond the wet lay-up. (b) A 2-in. band of thick bleeder material is placed just inside the molding compound. (c) The wet lay-up is covered with a flexible, clear, film diaphragm (poly(vinyl alcohol) film) and sealed to the mold with sealing compound. (d) The vacuum hose is connected to the mold. External pressure of 5 in. due to a vacuum inside the bag is slowly applied.

The wrinkles are worked out of the wet lay-up and film diaphragm. (e) Full vacuum is applied and then the air and excess resin removed from out of the lay-up into the bleeder material with a rubbing tool. (f) Full vacuum is maintained until the lay-up

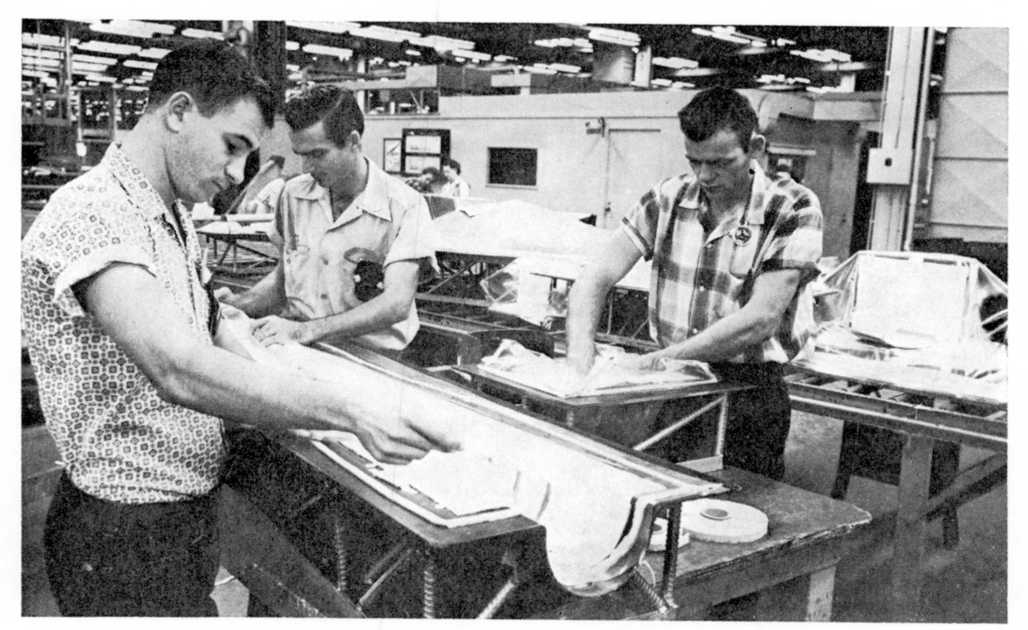

Fig. 8. Bagging. Courtesy General Dynamics Corp.

Fig. 9. Parts on dolly ready for curing. Courtesy General Dynamics Corp.

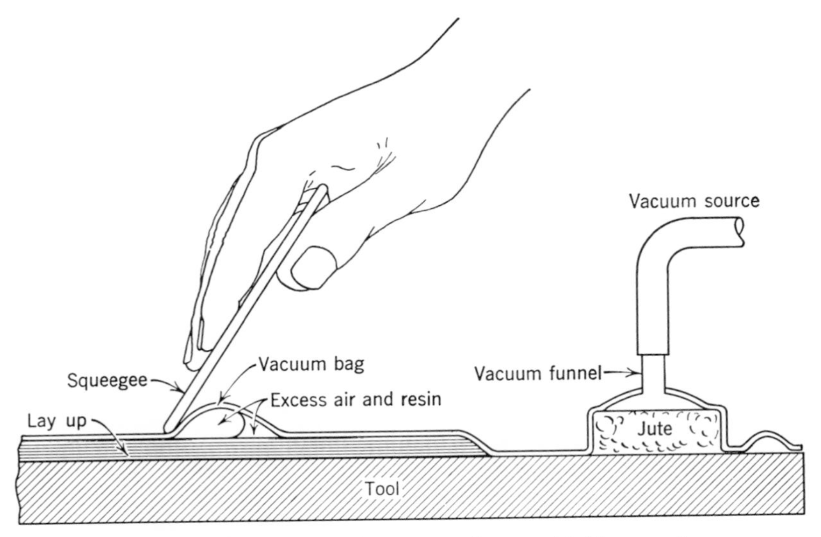

Fig. 10. Typical bagging system for a wet laid-up part.

has jelled hard. (g) When the part has jelled hard, it is placed on a portable curing rack ready for oven curing.

A typical bagging system for a wet lay-up part that is not required to be "void-free" is essentially the same as that for a preimpregnated lay up given above.

Area 5. Curing Oven and Autoclave. Generally speaking, most reinforced plastic parts made on good permanent tools can be cured either in an oven or an autoclave. The required curing pressure and the size of the available oven or autoclave facilities determines the method of cure. A cure cycle is a series of time–temperature steps at a given pressure, to transform an uncured reinforced plastic lay-up to a cured part that will meet contour and structural requirements.

The cure cycle of a part depends upon the type of resin system, the thickness of the glass-reinforced plastic part, the type of mold material, and the thickness of the mold. In order to maintain the assembly-line system, standard cure cycles are established for groups of similar parts (see Table 4).

The precure cycle transforms the resin from a liquid or semiliquid to a rigid solid. The post cure cycle continues the curing process to the point where the part will meet the required structural properties. Some standard cure cycles combine the precure and post cure cycles into one continuous cure cycle. The general steps in the curing operations are as follows: (a) All parts that have the same standard cure cycle are placed on the same cure dolly. (b) Full vacuum pressure is maintained on all parts. (c) When a full oven or autoclave load of parts has been collected, the curing processes are started. (d) The curing dollies are placed in the oven or autoclave, and the vacuum hoses are connected so that the vacuum pressure is registered on the gages outside of the curing chamber. (e) A final vacuum check is made on the parts to see if there are any leaks. (f) After the vacuum check has been passed, the doors are closed and the first step of the cure cycle begins. (g) For oven curing, the required vacuum is between 21 and 28 in. of mercury. (h) After the door of the autoclave has been closed, the air pressure and heat in the autoclave is increased, as required by the first step of the cure cycle. (i) After the required standard cure cycle has been completed, the air pressure

Table 4. Typical Cure Cycles for Structural Parts

Type of resin	Precure cycle		Postcure cycle		Approximate precure pressure, psi
	Temp, °F	Time, hr	Temp, °F	Time, hr	
general-purpose polyester	150	½	.....none.....		10–25
	200	1			
	250	½			
polyester modified with triallyl cyanurate	200	1	200	½	10–25
	250	1	300	½	
	300	½	350	1	
			450	1	
			500	2	
epoxy	150	1	250	½	10–50
	200	1	300	½	
	250	½	350	2	
phenolic	200	1	300	½	10–50
	250	½	350	2	
	275	½			

Table 5. Typical Mechanical Properties[a]

Type of test[b]	Temp, °F	General polyester	Modified polyester	Epoxy	Phenolic
ultimate compressive strength, psi	room temp	35,000	30,000	48,000	35,000
	160	27,000			
	300		25,000	25,000	26,000
ultimate tensile strength, psi	room temp	40,000	35,000	47,000	40,000
	160	31,000			
	300		32,000	30,000	30,000
ultimate flexural strength, psi	room temp	50,000	48,000	65,000	50,000
	160	40,000			
	300		40,000	42,000	40,000

[a] Tested after 30-min exposure at temperature.
[b] Using 181 glass fabric tested in the warp direction.

in the autoclave is released. The door of the autoclave and the ovens are opened slightly to reduce the time required to cool the part. (j) All parts are cooled uniformly to below 150°F under vacuum pressure to minimize thermal shock and warpage. (k) Parts are now ready to be moved to the part-removal area.

Area 6. Part Removal, Trim, and Drill. After the parts have been removed from the oven or autoclave, they are taken to the part-removal area. The bagging materials and parts are carefully removed from the molds. Any scratches or dents to a mold surface will be reproduced in the next molded part (see Fig. 11). The molds go back to the mold-storage area to be cleaned and prepared for the next molding cycle. The parts are taken to the trim and drill room. Each production part has its corresponding trim shell, which is made to the exact outside perimeter of the finished part. The trim shell is put over the part and the finished trim line is scribed onto the part. The trim shell also contains drill bushing, for accurately locating all the holes that have to be drilled in the part. With the trim shell properly located, the proper size holes are drilled in the part. The trim shell is removed and the excess material is

Fig. 11. Removing parts from molds after curing. Courtesy General Dynamics Corp.

removed by a sawing or routing operation. The edge of the part is given a light sanding and the part is ready for final inspection.

Area 7. Final Inspection. Each part is given a thorough inspection (see Fig. 12) to make sure that it meets the requirements of the blueprint and any other applicable documents. Typical defects that should be looked for are "starved" areas, fractures, voids, delaminations, blisters, and resin pockets. Typical critical areas that should be checked are the thickness of flanges, size and location of holes, overall dimensions, and general surface conditions.

If a part has the critical structural requirement, it may have a test tab molded with it, which serves as a quality-control sample for that part. The test tab will be sent to a quality-control test laboratory and tested. The test tab must pass the required structural values as a part of the final inspection requirements for the part it represents (see Fig. 13). When the part passes final inspection, it is ready to be routed to its next destination as a completed, reinforced plastic part, produced by the bag-molding process.

Fig. 12. Final inspection. Courtesy General Dynamics Corp.

General Design Information

It is important for the modern designer to understand the flexibility of designs that are possible because of the versatility of reinforced plastic materials and the bag-molding processes. The most common materials used are listed in Table 1, and typical structural properties of these materials are shown in Table 5.

Miscellaneous Design Rules

1. The minimum mold radius is $\frac{3}{16}$ in.
2. The minimum draft angle is 2 degrees.
3. Undercuts can be made, but they increase the tooling cost.
4. Only one surface of the part can be controlled by the mold surface.
5. Local build-up areas can be made where desired, but whenever possible locate them on the bag side.
6. Metal inserts of aluminum or steel may be used if small. Large aluminum inserts may cause trouble because of the difference in thermal expansion.
7. The limiting size of the part is determined by curing facilities.
8. Semitransparent or translucent parts can be made by selecting the correct resin system.
9. Strength orientation will be determined by the type of reinforcement used, and how it is laid up.

Fig. 13. Quality-control testing. Courtesy General Dynamics Corp.

10. Normal thickness variations depend upon the thickness uniformity of the reinforcement, and the number of plies of reinforcement. As a typical example, for pre-impregnated material, phenolic, style 181 glass fabric, the variations are as follows:

Blueprint call-out, in.	*Tolerances, in.*
0.015–0.060	±0.010
0.061–0.119	±0.015
0.120–0.249	±0.020
0.250–0.380	±0.025
above 0.380	±0.030

11. Typical resin-to-glass ratios are 50% glass mat, 50% resin; 64% glass fabric. 36% resin.

Bibliography

1. R. H. Sonneborn, *Fiberglas Reinforced Plastics,* Reinhold Publishing Corp., New York, 1954.
2. *Modern Plastic Encyclopedia,* 1963.
3. "Reinforced Plastics, The New Material of Construction," *Mater. Design Eng.* (Feb. 1961)
4. "MIL-HDBK-17 Plastics for Flight Vehicle," *Reinforced Plastics* Part I, 7–61 (Nov. 1959).
5. *Fiberglas Reinforced Plastics,* Owens-Corning Fiberglas Corp., Aug. 1962.

Sam Monroe
General Dynamics Corporation

MOLDS

The molds used for the numerous types of plastic and the several types of molding methods in commercial use constitute a controlling factor in the molding industry, the most advantageous utilization of both plastic and molding equipment requiring a properly designed and constructed mold. This article will describe molds used for compression and transfer molding of thermosetting polymers and for injection molding of thermoplastics. Molds for blow molding are described in the article on Blow Molding, under MOLDING, in conjunction with the machinery of which they are a part in this type of operation; rotational molds are similarly discussed in Rotational Molding under this same title. Discussions of the processes of compression, transfer, and injection molding, as well as a glossary of terms used in conjunction with molds and molding, are also given under MOLDING.

The mold consists basically of a cavity and a plunger, or force. Most molds also include an ejection mechanism for removing the molded article; transfer and injection molds also commonly include systems of runners and gates through which the molten plastic enters the mold. These and other components of a sophisticated modern mold are illustrated in the figures presented throughout this article.

Types of Molds

Molds can be divided into four general classes: prototype, hand, semiautomatic and automatic molds. The best type to be used depends on the product design, molding material, equipment available, production requirements, and economic factors.

Prototype molds are used when a limited number of items is to be produced, as, for example, for the evaluation of a new product. These molds are made from easily machined metals or from casting materials, and the mold details are assembled into supporting frames.

Hand molds are the oldest type of mold, and can be constructed in either simple or sophisticated designs. They must be removed from the press for loading and unloading and are used for short production runs or for designs requiring many inserts and complicated coring geometry. Use of hand molds results in higher labor costs, lower production rates, and thus higher unit costs than those for the use of semiautomatic or automatic molds.

Semiautomatic molds are mounted into a press for the duration of a production run. The molded part is automatically ejected from the cavity as the press opens, but an operator is required to load the molding material, actuate press controls for the molding sequence, and remove the piece from the mold. This method is widely used for all types of molding.

Automatic molds include additional mechanical components such that, when the mold is mounted in press, the complete operation sequence from the time the material is loaded until the completed parts are discharged from the mold is carried out automatically. Automatic molds offer the most economical approach to long production runs, because labor costs are kept to a minimum and all molding operations are accurately controlled. Thermoplastics can be molded at rates up to 700 cycles/hr, and thermosetting polymers at up to 450 cycles/hr; tooling must be of the highest quality to meet the exacting demands of high-speed production.

Factors in Mold Design

The proper design of a mold must take into consideration the molding material, the end use of the part, and requirements regarding appearance and tolerances. These will offer guidelines for the type of mold finish, dimensional accuracy of the mold, parting-line delineation, locations of ejector pins and of gating, and type of processing. Reviews of the following factors are also helpful in the design of a mold: (a) tapers, tool strength, ejector strength, undercuts, insert locations, proper seal (to prevent flashing), and general cross section of the design for optimum physical properties of the molded article and economical use of equipment and material; (b) mating parts or auxiliary operations for possible incorporation into a one-part design for additional savings; (c) quality and finishing standards which must be met by the product and mold design; (d) troublesome phases in the molding process which must be planned for ease of mold maintenance. Finally, a designer must have a good working knowledge of the characteristics of the molding material and the process to be used.

Gating. Gates, in a transfer or injection mold, must be correctly located to permit proper flow of material after it enters the mold and easy removal of the gate after molding. Poor gating leads to uneconomical molding cycles and, in many cases, weakens the finished part through flow-induced orientation of the polymer (see also CRAZING).

Venting. Provision must be made in all mold designs for the escape of air while the molding material is entering the cavity and of gases produced during curing. Parting-line areas are vented by machining grooves 0.001–0.003 in. deep and of various widths (typically 0.062–0.250 in.), depending on the overall size of the product. Caution must be exercised to ensure that the system permits passage to the atmosphere when the mold is in a closed position. Ejector pins, sleeve ejectors, and other moving members in the mold have clearances incorporated to allow for venting. Moldings having deep pockets, ribs, or projections require venting in these areas by ejectors, with venting grooves machined on the pins or on the clearance areas in the surrounding mold.

Improper venting, especially in modern high-speed molding, is probably the most serious source of trouble in mold designing. This is always a critical area in all mold designs, whether it be a hand mold or a highly sophisticated automatic design. Air or trapped gas in a cavity does not permit the flow of the molding material to all areas, and this results in porosity, unfilled parts, poor finish, and dimensional inconsistency.

Ejection. Removal of a finished molding from a mold is conventionally accomplished by a series of round ejector pins assembled into a pin plate fastened to an ejector bar. This last is actuated by the press stroke and travels to a predetermined position. The round ejector pins are located so as to engage on the molding surface at the parting line, and, if necessary, on the interior of the part. Larger areas of contact and short distance of movement are preferable. If the ejector does not travel far enough for the part to clear the molding members lowered efficiency and higher operating costs result. Use of ejectors that are too small in area results in mold breakage. The molding at the time of press opening may not be as rigid as after it has reached room temperature; for this reason ejectors of maximum size should be used on all areas where the molding may adhere to the mold. Ordinarily, all ejection systems have positive return pins incorporated for accurate pre-positioning of the assembly prior to the next molding cycle.

Designs not permitting round ejectors may employ blade or shaped members to fit a specific design. Enlarged, rectangular-shaped blade ejectors may be used where internal ribbing and small cross sections make conventional round pins unsuitable. Molding designs with internal bosses that need center core pins require *bushings* or *sleeves* acting as ejectors to provide uniform movement from the mold. Valve ejection permits a large contact area for the ejector on an internal surface of the molding. Good tool strength is obtained since the ejector pin is machined in the shape of a valve stem.

(**a**)

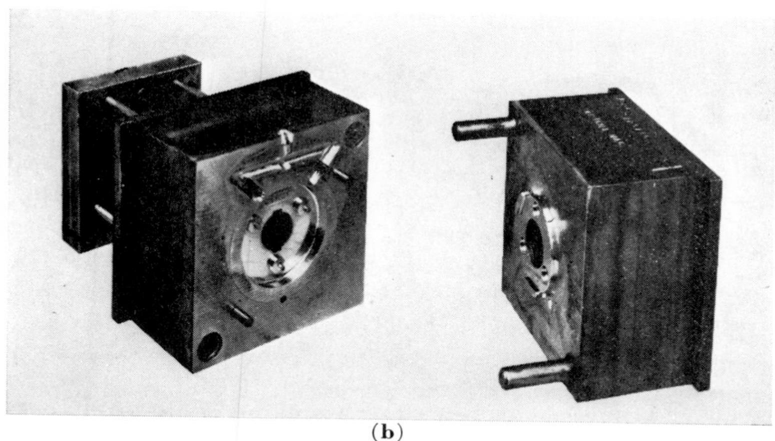

(**b**)

Fig. 1. (**a**) Standard transfer frame for plunger transfer molding: A, mounting plate; B, positioning clamp for holding mold unit; C, cartridge heater hole; D, nesting area for mold unit; E, transfer chamber; F, leader pin; G, leader pin bushing; H, transfer cull block; I, transfer runner; J, cartridge heater hole; K, nesting area for mold unit; L, positioning clamp for holding mold unit; M, grid; N, ejector bar; O, mounting plate. (**b**) Master unit transfer mold insert. Courtesy Master Unit Die Products, Inc.

Fig. 2. Standard injection frame. Courtesy Master Unit Die Products, Inc.

Air ejection is used for moldings requiring rapid and positive removal from the mold. High-pressure air enters through a port and actuates a valve-type ejector pin. Normally, the valve, upon completion of the ejection stroke, is returned to the molding position by a spring.

Moldings with deep internal sections or undercuts, which would not be freely ejected in a conventional system, require *two-stage ejection*. One system will remove the internal mold sections, and the second stage will free the part from the mold.

Channeling. Channels may be required for circulation of heating or cooling media through a mold. Molds for thermosetting plastics may be heated by steam or other medium in this manner. Electrical cartridge or strip heater bands are also universally used for mold heating. Proper channeling on molds designed for thermoplastics permits the removal of heat from the material in a closed mold. It is essential for cooling to be incorporated in all mold sections, and for the mold design to permit the greatest feasible cooling. (Some thermoplastic polymers require heating in the mold; in this case the same principles apply.)

Multiple-Tray Designs. Moldings requiring many inserts, or special cores, in multicavity molds are economically produced in molds using duplicate cavity or plunger trays to permit the unloading and reloading of the special cores or inserts, while the second tray is in the press cycling. This permits most efficient use of the press cycle, with the extra trays prepared for immediate insertion into the press upon the removal of the moldings. Specially designed automatic insert fixtures may be developed for high-volume production requirements for multiple-tray operation, or, if economics merit, for direct loading into the mold.

Standard Frames. Many types of standard mold frames have been developed by molding companies and commercial tool companies (Figs. 1–4). The frame is mounted in the press and contains provisions for predetermined mold units with a master ejection assembly and sprue arrangement. The units contain an auxiliary ejection mechanism and are inserted into the master frame. Individual units may be employed or as many as four units may be installed in the frame for production. Molding material and color requirement must be compatible for all units that are used at

Fig. 3. Standard injection frame, with individual mold units assembled into the frame before fabrication. Courtesy Master Unit Die Products, Inc.

Fig. 4. Standard mold frame mold unit, showing injection plastic moldings connected to runner systems. Courtesy Master Unit Die Products, Inc.

one time. Preparation time is kept to a minimum for mold construction, mold setup, and mold investment, and a minimum of mold storage area is required.

Dimensional Tolerances. Reliable dimensional accuracy in the mold permits moldings to be produced economically within the specifications required for the part. Designs demanding tolerances beyond the capability of molding technology require auxiliary operations, such as machining (qv), drilling, reaming, or sizing to very close limits.

Installation and Maintenance. Setting up or installing the mold in a press must be planned so that the mold will physically fit into the equipment and so that ejection or moving members of the mold will be compatible with the mechanism of the press. Gross volume or weight of the molding charge must be within the press capacities. Projected areas of the moldings, and, in transfer and injection molding, the sprue, cull, and runner systems must be included. These total areas must be compatible with the press capacity and type of molding material specified to permit proper molding pressures.

Mold sections, core or ejection pins, susceptible to wear or breakage during production must be designed for ease of replacement and of disassembly and reassembly of the mold. Operating specifications covering mold assembly, materials and processing, finishing, and quality control must be issued.

Molds for Thermosetting Polymers

Thermosetting polymers are commonly molded by either compression or transfer molding. In *compression molding,* molding material of a specified weight or volume, as either loose powder or preform, is placed in a displacement cavity; single- or multiple-cavity molds may be used. The press has a stationary platen and a moving platen actuated to close the mold and to maintain pressure and heat on the molding material for the time necessary to form and cure the part. Excess material flows out of the cavity at the parting line of the mold to form flash, which must be removed by auxiliary operations.

In *transfer molding,* the molding material is preformed to a particular size and then placed in a chamber or tube from which it is moved through a runner and gate system into the closed cavity and plunger sections of the mold. This method makes possible the molding of fragile and delicate designs, inserts, and deep holes of small diameters. Single- or multiple-cavity molds may be employed. Ordinarily, parts made in this process have little or no flash at the parting lines. Cycling is more efficient, tool life is longer (except in areas where high wear is caused by abrasion through runners and gates), and fewer auxiliary operations for hole forming are required than in compression molding. In most cases, a degating operation is necessary.

Generally, the material required for a given number of parts is greater than for compression molding owing to the transfer runner system. However, in most cases, this is offset by greater molding efficiencies; in other cases, the product design could not be realized except by this method because of the high costs of operations that would be required if compression molding were used. Ordinarily, values of physical properties furnished by suppliers are those of compression-molded specimens; these values are often lower for transfer moldings. Material suppliers' recommendations should be consulted on this point in designing equipment for transfer molding.

Equipment available for transfer molding includes: (a) Hand transfer. (b) Integral floating-pot transfer. In this method conventional single-movement presses may be used with the mold designed for transfer molding. (c) Plunger, or top-ram transfer. This molding equipment has a main cylinder to actuate the mold opening and closing, and an independent cylinder for forward and return strokes of the transfer plunger to move the molding material from the transfer chamber through the runner and gating system into the mold cavities. There are many variations of this equipment in which the plunger cylinder may be mounted at the top or bottom of the press; in some cases horizontal plunger cylinders may be employed for special applications.

Some new designs for transfer-molding equipment utilize some characteristics of injection molding, in that the preheated molding material is injected into a clamped mold. By such methods faster cycles with resultant economic advantages can be achieved. Reciprocating-screw transfer molding utilizes a two-stage press in which a reciprocating screw is used to meter material into a conventional bottom-ram transfer chamber for movement into the mold cavities. Curing cycles are reduced and, in addition, the molded parts and runner system are separated in a fully automatic operation. This system eliminates preforming of the material, preheating, and excessive material and parts handling.

The reciprocating-screw design is basically the same as the design used in injection molding of thermoplastic materials. The material enters the mold through a sprue, with a runner and gate system being used for multiple-cavity molds. This system avoids the culls formed in other types of transfer molding. The plunger injection system is similar to the system used in conventional plunger injection molding, in which the molding material is placed in a hopper and fed by gravity into specially constructed heating cylinders. It is then forced by a double-acting plunger through a sprue nozzle into gated cavities. These processes have a distinct advantage over other types of transfer molding in that they do not require preforming, separate preheating of the molding material, or high costs of direct labor and material handling. In most applications the equipment needed may be set up and used automatically.

Operation may be hand, semiautomatic, or fully automatic for both compression and transfer molding. The *hand*-mold components are placed in the press without any permanent mounting, and, upon completion of the curing cycle, are removed from the press. Separation of the part from the mold is accomplished away from the press.

Semiautomatic molds are, and no doubt will remain, the mainstay for the many molded thermosetting products manufactured. The mold is mounted into a press, one half becoming the stationary half by being fastened to the permanent platen, and the other half becoming the moving half of the mold. The latter is fastened to the press

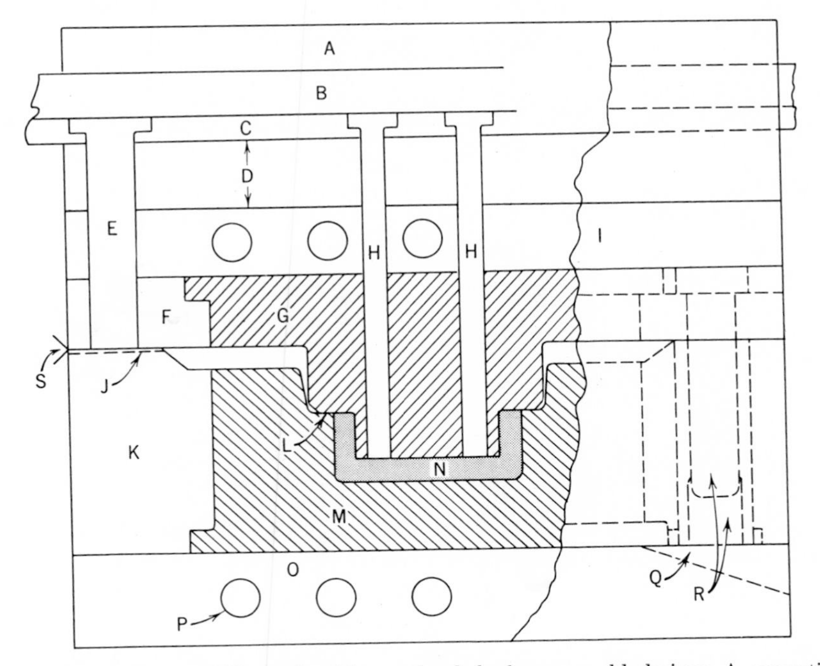

Fig. 5. Compression mold, semipositive or landed plunger mold design: A, mounting plate; B, ejector bar; C, pin plate; D, ejector travel; E, return pin; F, retainer; G, plunger; H, ejector pins; I, support plate; J, vent; K, cavity retainer plate; L, parting line; M, cavity; N, molded part; O, mounting plate; P, heating channels; Q, clearance; R, leader pin and bushing; S, area of mold opening.

platen, which is actuated by the press ram or other mechanical operation to obtain a mold closing and opening sequence. Ejection of the part from the cavity is accomplished through a series of ejectors fastened to an ejector plate and bar which engages an ejection system on the press toward the end of the mold-opening stroke. *Automatic* molds may be used in conventional transfer and in reciprocating-screw equipment.

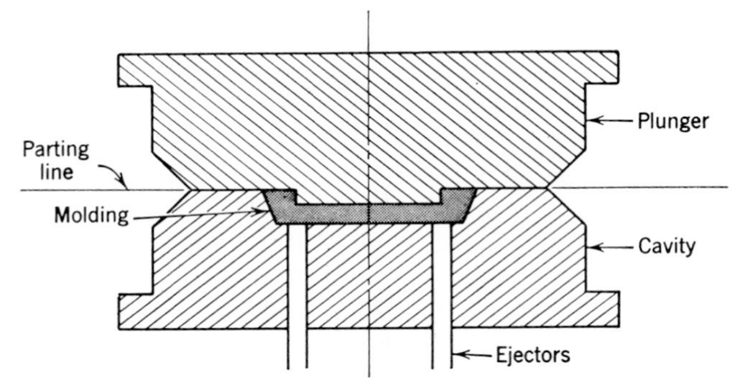

Fig. 6. Flash mold.

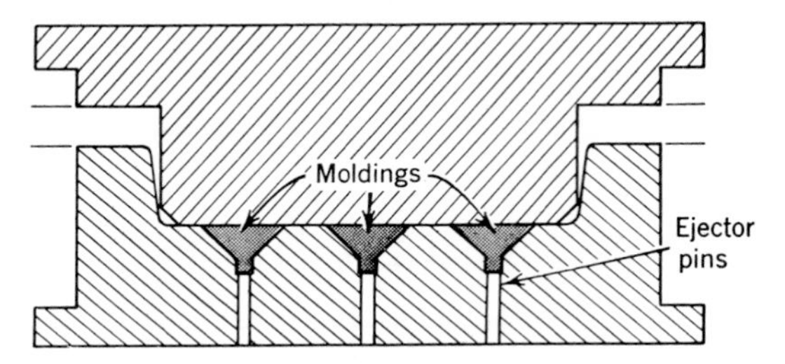

Fig. 7. Gang or sub-cavity mold.

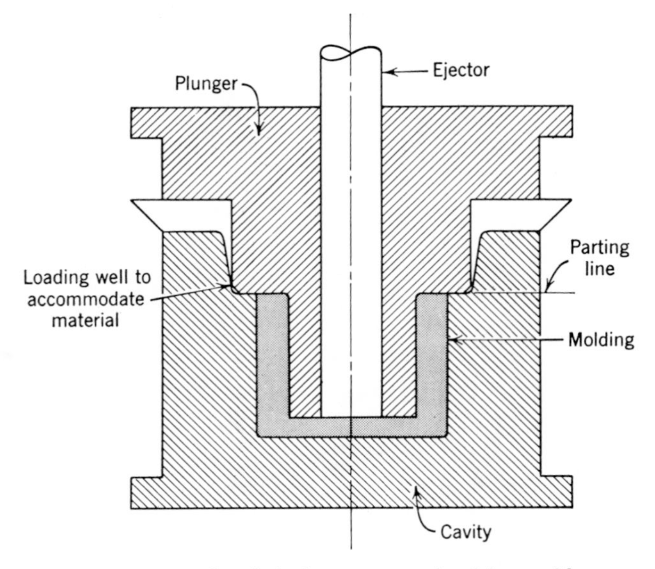

Fig. 8. Landed plunger or semipositive mold.

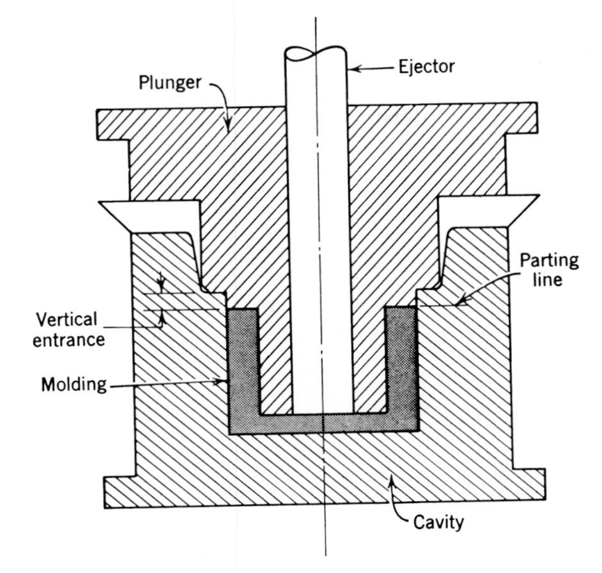

Fig. 9. Landed plunger or semipositive mold with vertical positive entrance.

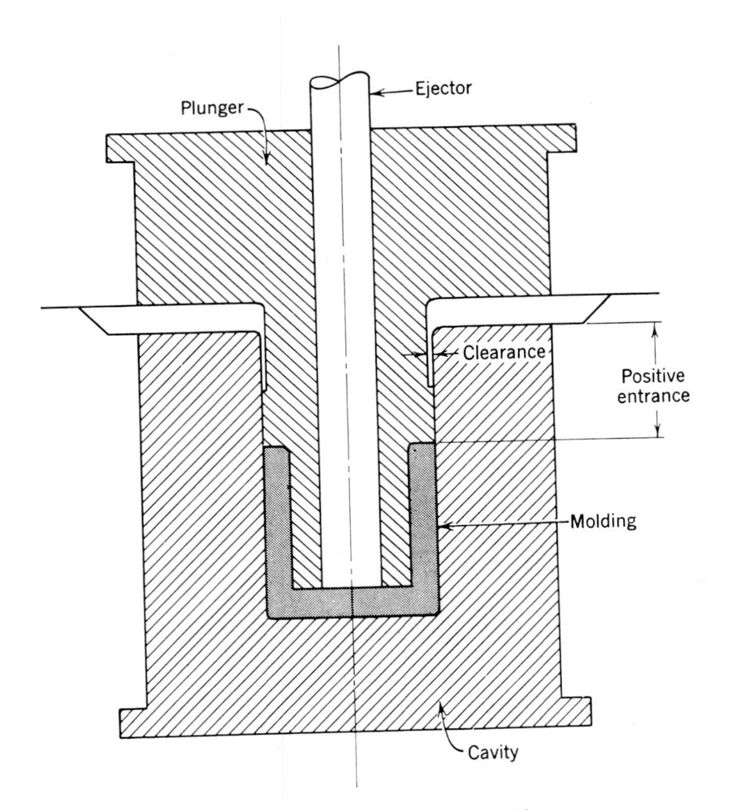

Fig. 10. A typical positive mold.

Compression Molds. There are several basic types of mold designs in use for compression molding. These will be discussed in this section. A typical compression mold is shown in Figure 5.

In *flash* molds (Fig. 6) a charge of predetermined weight or volume is placed directly into the mold cavity and the molding formed under pressure and heat. Excess

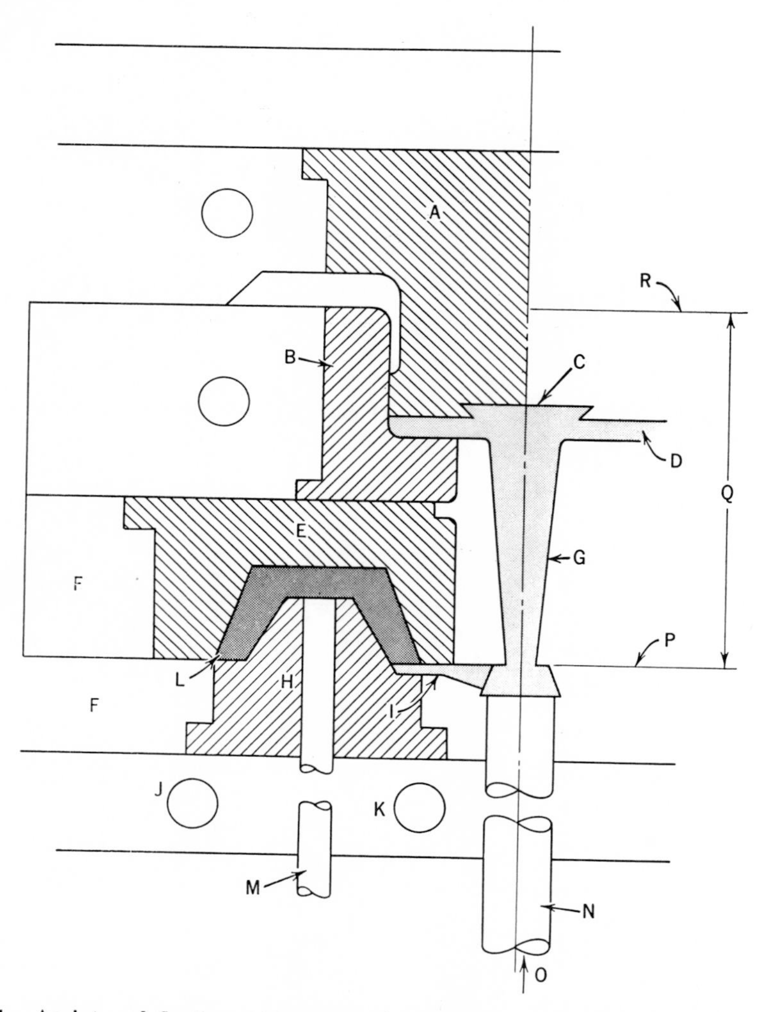

Fig. 11. An integral floating-pot-type transfer mold: A, transfer plunger; B, transfer pot; C, cull pickup; D, cull; E, cavity; F, retainer; G, sprue; H, plunger; I, gate; J, support plate; K, heating channels; L, molded part; M, part ejector; N, sprue ejector; O, press movement; P, mold parting line; Q, floating transfer pot assembly; R, mold opening for loading material.

material flows horizontally away from the closed cavity into clearance areas between the stationary and moving halves of the mold. The excess material, or flash, requires finishing operations to remove it from the molded part. Single- or multiple-cavity molds may be used. This design may be used for all general types of molding materials and for some modified types.

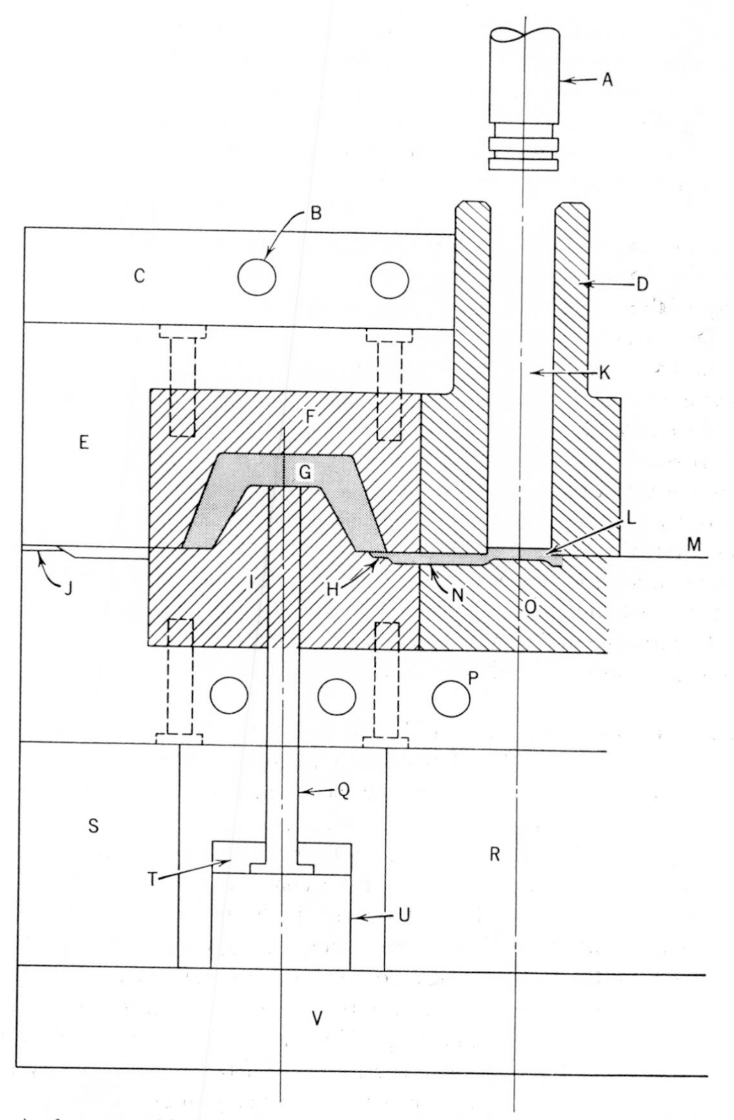

Fig. 12. A plunger mold: A, transfer plunger (double-acting stroke); B, heating channels; C, top plate; D, transfer tube; E, retainer; F, cavity; G, molding; H, gate; I, plunger; J, vent; K, loading chamber to accommodate material with ample depth to prevent material from escaping at the top during plunger movement; L, cull; M, parting line; N, runner; O, cull block; P, retainer; Q, ejector pin; R, support; S, support grid; T, pin plate; U, ejector bar; V, bottom plate.

Miniature parts are generally molded in *gang or sub-cavity molds* (Fig. 7). These permit a cluster of parts to be made with a minimum of flash, since the parts can be grouped closely together and one material charge used for the many parts. Most molding materials, with the exception of those with high bulk factors, can be used in this design. This design requires a landed plunger or semipositive mold.

The *semipositive or landed plunger* is probably the most universally applied mold design; it is used for nearly all thermosetting plastics. These molds differ from the flash type by incorporating a well above the molding part, with a landed area extending 1/8–1/4 in. beyond the periphery of the molded part (Fig. 8). The well must be deep enough to accommodate the required amount of molding material as preform or powder. Excess material flows horizontally away from the part at the parting line, through a clearance area in the plunger well. This design permits better indexing of cavities and plungers to give moldings with more uniform cross sections, and also allows for material displacement.

The *semipositive or landed plunger with a vertical positive entrance* differs from the preceding design by an extra vertical section incorporated on the plunger, formed to the periphery of the cavity and extending into the cavity (ordinarily from 1/16 to 1/8 in.) (Fig. 9). The cavity depth must correspond. This mold produces a vertical flash at the top of the molding for designs that require this type of finishing. It is used for almost all types of thermosetting materials, and, in particular, for large moldings.

In *positive molds* the entire pressure is in the cavity, whereas in semipositive molds part of the pressure is on the land. The plunger has a vertical area conforming to the internal shape of the cavity (Fig. 10). The height of the vertical entrance and succeeding depth into the cavity proper are determined by the displacement necessary for the molding material plus a nominal length for guidance. The molding charge must be accurately controlled for dimensionally accurate parts, since the weight of the charge determines the finished size. This design produces a vertical flash on the molded part and allows all available molding pressure to be applied on the molding. It yields dense moldings, and is generally used on reinforced filled materials or on those with a high bulk factor.

Transfer Molds. Transfer molds are usually of the flash type since no extra loading space is required. Two basic mold arrangements are utilized for transfer molding, with modifications to suit a particular molding or equipment. The older of the two is *integral,* or pot-type, transfer. The mold cavities and plungers are similar to those of a flash mold, but include gates, runners, and transfer chamber (Fig. 11). The transfer chamber is above the cavity. The area of the transfer chamber normally must exceed the total cavity area (including sprues and runners) by at least 25% to ensure that the mold does not open during the molding cycle. The depth of the chamber is based on the total volume of molding material required plus an additional height of the chamber above the top of the material to prevent this material from escaping at the time of mold closing. The material flows into the cavity through tapered sprues or through a runner and gate system to multiple-cavity molds. This design requires a sprue puller in the transfer-chamber plunger in order to remove the cull, and the sprue extensions.

Top- or bottom-ram transfer molds, also called *plunger molds,* are the most common design (Fig. 12). This equipment is used in methods designed for a particular use, whereas integral-transfer molds are designed to operate in standard molding presses. The center of the plunger mold generally has an open tube for the material loading chamber, with a plunger entering from the top or bottom to move the material from the chamber into the runner system, gates, and mold cavities. These molds require a careful check of the pressures necessary for a given combination of material, cavity, runner and chamber area, press-clamping capacity, transfer-ram capacity, and available pressure for the transfer plunger. Normally, the basic equipment covers a 4–5 to 1 ratio for clamping and transfer pressures, eg, a 100-ton clamping ram has a 20–25 ton transfer ram. The molding pressure would be developed from these and would

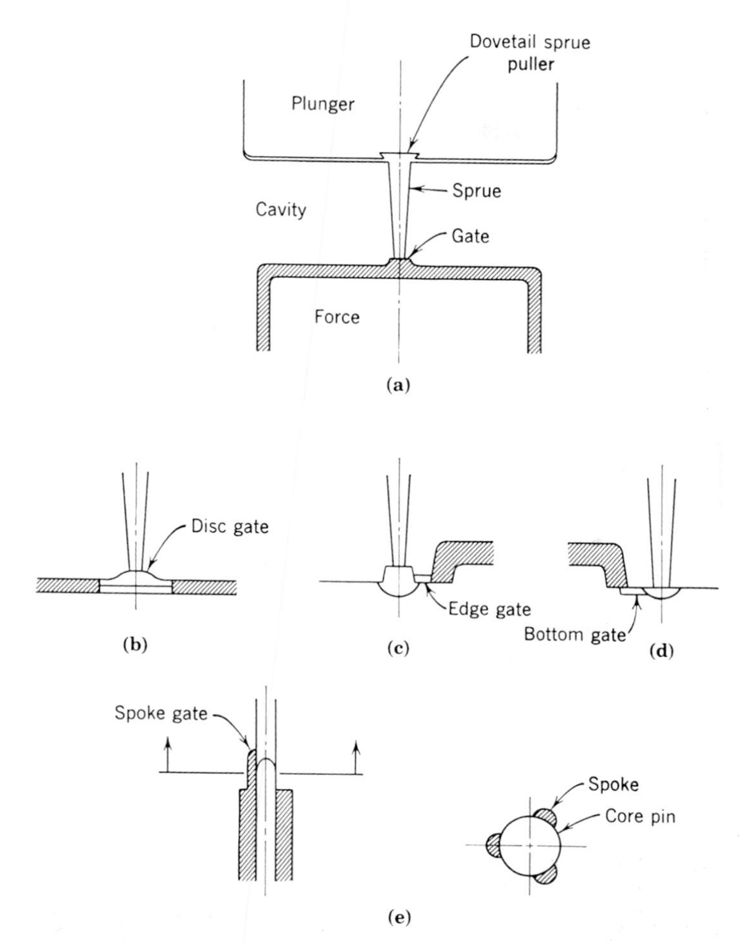

Fig. 13. Gate designs used in integral transfer molds. The molded article is in the crosshatched area.

vary depending upon the transfer plunger employed. In all cases, the clamping pressure must exceed the transfer pressure by a safety margin wide enough to ensure that the mold does not open during the transfer cycle.

Breaking-in procedures and maintenance of transfer molds are discussed in the article on Compression and Transfer Molding under MOLDING.

Design Variations. Special types of products and sophisticated designs may necessitate design variations to permit economical production. Some of the most important of these will be described in this section.

Split Cavity or Split Wedge. Pieces with projections which prevent removal by a straight withdrawal can be molded in split-cavity molds. The entire cavity section is removed from the mold as a unit and then separated for removal of the molded part. A split wedge can also be used for parts with undercuts or projections opposing conventional opening of the press (mold sections and other loose parts being generally called wedges). This permits the molding and the corresponding metal section of the mold forming the undercuts or projections to be removed together from the mold by the

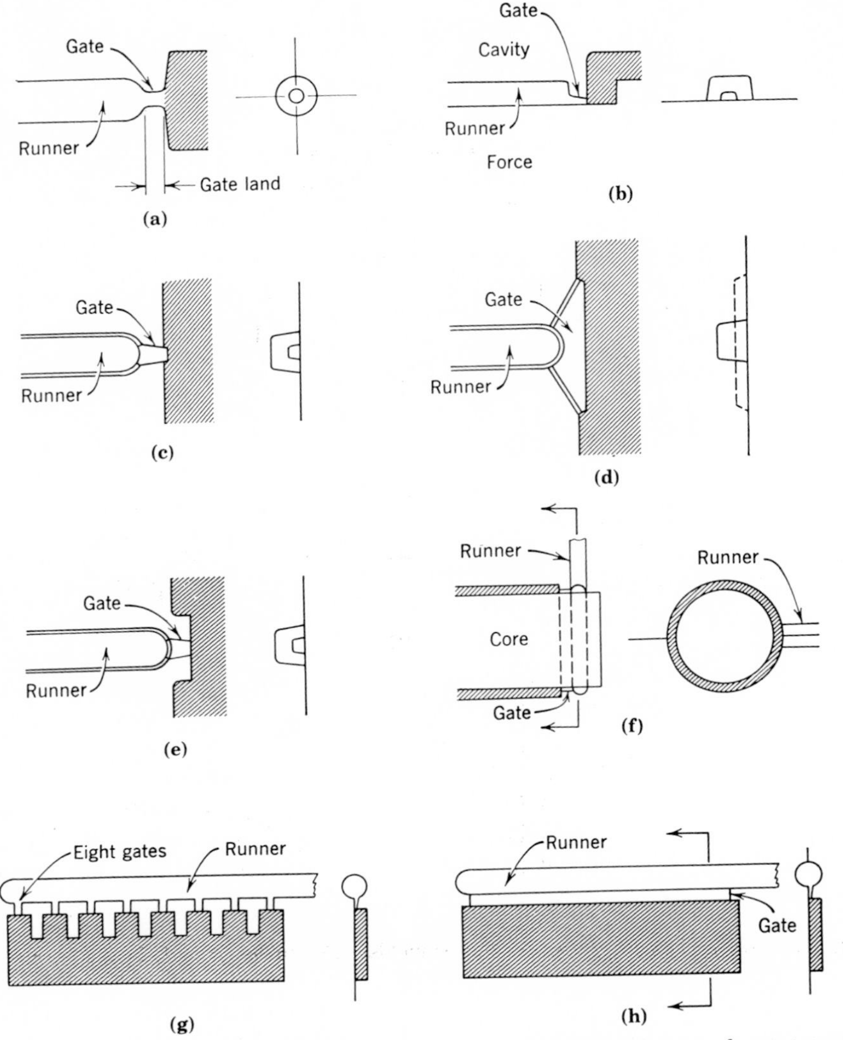

Fig. 14. Gate designs used in plunger transfer molds: (**a**) edge gate, round runner and gate; (**b**) edge gate, trapezoidal runner; (**c**) flat gate (edge or bottom); (**d**) fan gate (edge or bottom); (**e**) depressed edge gate; (**f**) ring gate; (**g**) multi-gate (edge); (**h**) flash gate. The molded article is in the crosshatched area.

press stroke and opening. The finished part is removed from the surrounding metal sections by hand or auxiliary fixtures. Efficient operation requires a minimum of two sets of the wedge details, one set for part removal at the time the second set is installed in the mold and undergoing the curing cycle. The outer shape of the wedge is normally made in two parts, for separation. The external area is tapered for positioning in the mold frame. Wedge halves are interlocked, and, in order to insure proper assembly and to prevent misalignment, must be carefully indexed.

Loose Details. Moldings with horizontal or angular surfaces are formed by separate mold details which are removed from the completed molding. Normally, for economical molding, duplicate sets are used.

Stripper Plate. Moldings with thin cross sections, minimum wall sections, and deep taper require uniform ejection. The molding area does not permit the use of conventional ejector pins. Instead, a movable plate located above the cavity and actuated by the press opening is used. As the press opens, the stripper plate moves down, using the top surface of the molding as the ejection area and stripping the molding from the plunger. Sometimes, if design permits, conventional ejectors should be employed in conjunction with a stripper plate.

Spring Box. Moldings in which fragile mold sections are exposed in the cavity or which have long inserts, and which cannot be transfer molded, require protection of these components during the initial phase of press closing and flow of material. This is accomplished by adding a spring assembly to the mold base. The mold is initially closed with a spacer fork inserted to restrict the closing of the press to approximately one-third to one-half of the total depth. Shortly after initial mold closing and when the material is in a fluid stage, the mold is opened slightly and the fork removed to allow complete closure.

Automatic Unscrewing. Parts with internal or external threads may be freed from the mold by means of a rack and gear mechanism or of mold wipers sequentially timed with the press movement.

Cam and Hydraulically Operated Cores. Moldings with horizontal or angular surfaces that are to be produced in volume can be efficiently molded with mold designs incorporating members operated either hydraulically or by a mechanical cam. Cam movements are actuated by press movement, and hydraulic designs are actuated by valving from the press power system.

Heating. Two methods for heating molds are in general use. In one, steam at a predetermined pressure and temperature is circulated through a series of channels machined into the mold plates (and in individual cavities and plungers if required). The other method is electrical; heating cartridges are inserted into mold base channels (and cavity and plunger if required) and the temperature thermostatically controlled.

Gating. Integral transfer-mold designs normally employ the types of gate designs shown in Figure 13, depending upon the design of the molded part. Plunger transfer-mold designs normally employ the types of gate designs shown in Figure 14. Typical variations of runner systems are also shown in the figures.

Molds for Thermoplastics

Thermoplastic materials are commonly molded by injection-molding methods In *plunger* injection molding, material is fed from a hopper into a heating chamber where a double-acting plunger forces the material from the cylinder through a nozzle and a system of runners and gates into the cavity. Single-cavity molds eliminate the runner system.

The first stage of *two-stage plunger* injection is similar to single-stage plunger injection except that instead of flowing directly into the mold, the material flows into a second-stage chamber, normally of larger capacity, from which a double-acting plunger transfers the material into the mold. This system improves plasticizing of the material, improves control of metered weight, and makes possible a machine of larger capacity.

The *reciprocating-screw* system employs a screw to mix and melt the material in a cylinder. After the material is melted it builds up in front of the screw tip, forcing the

Fig. 15. A vertical reciprocating-screw injection-molding press with a 3-oz, 50-ton clamp. Courtesy Stokes Equipment Division, Pennsalt Chemical Corp.

screw to retract; the revolving motion stops and the screw becomes a plunger with a forward motion which moves the material into the mold (Fig. 15).

Normally, the *two-stage screw press* employs a fixed screw to plasticize the material, forcing it into a second chamber from which it is transferred into the mold by a double-acting plunger. Equipment is available with horizontal or vertical opening, and in many modifications for custom specialty molding.

Hand, semiautomatic, or automatic molds may be used in injection molding. *Hand molds* are handled by an operator and are placed into an injection press for molding. Upon completion of the molding cycle, the mold is removed from the press and separated outside the press to remove the molded part. Modified hand molds permit cavity and plunger holders to be mounted to the press, with mold inserts employed for the specific design. The inserts and molding are removed from the press as an assembly, and disassembled by an operator. *Semiautomatic molds* are fastened to the fixed and moving platens of the press and are used for all types of injection molding.

Automatic molds require positive ejection of the parts from the mold and positive release of supplementary runner and sprue systems. A low-pressure closing is necessary to prevent damage to the mold by improperly ejected parts. Hoppers should be equipped with magnets to prevent metal bits from entering the heating chambers. Extreme caution in mold design is necessary to ensure proper operation; it is often necessary to employ air ejection systems, auxiliary wiper controls to sweep the parts, and

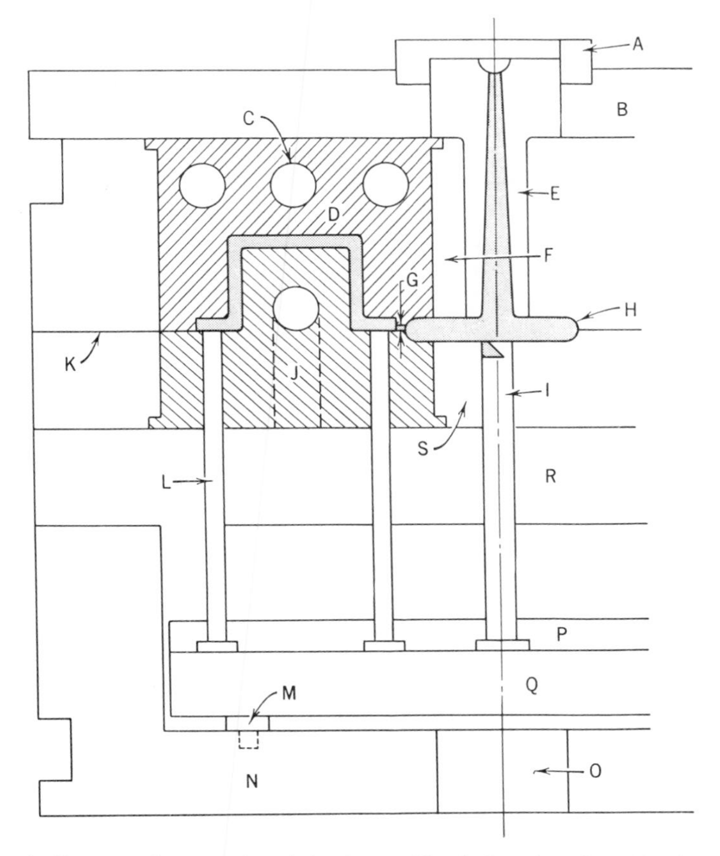

Fig. 16. A diagram of a two-plate injection mold: A, locating ring; B, clamping plate; C, water channels; D, cavity; E, sprue bushing; F, cavity retainer; G, gate; H, full round runner; I, sprue puller pin; J, plunger; K, parting line; L, ejector pin; M, stop pin; N, ejector housing; O, press ejector clearance; P, pin plate; Q, ejector bar; R, support plate; S, plunger retainer.

runner systems from the mold faces. Many modifications are in use for automatic operations, eg, three-plate molding, which permits separation of the parts from the runner system. Conveyors may be used to handle parts and to remove runners and sprues into material grinders and blenders for feeding back into the hopper. Hot-runner and insulated-runner systems facilitate the flow of the molten material to the cavities to give the highest efficiency of material utilization.

All these modifactions and, if necessary, special equipment and accessories may be designed for insert loading, and are used for high-volume production.

Mold Designs. Molds used for thermoplastic polymers are commonly flash molds, since in injection molding, as in transfer molding, no extra loading space is required. However, the wide variety of thermoplastic materials which are injection molded requires a large number of variations on this basic type. The ones in common use are discussed in this section.

The *two-plate* design is the one most commonly used for all types of materials. The cavities are assembled into one plate and the plungers installed into the second plate. The sprue bushing is set into the plate, which is mounted to the stationary half of the mold; this arrangement permits the use of a direct center gate leading either into a single-cavity mold or into a runner system for a multicavity mold. The moving half of the mold normally contains the plungers and ejector assembly and, in most cases, the runner system. This is the basic design for an injection mold (Fig. 16), all other designs having been developed to meet special requirements.

In a *three-plate* mold a third, movable, plate is added to the design (Fig. 17). The cavities are located in this plate, permitting center or offset gating into each cavity for multicavity operation. Upon opening, the mold separates in two openings, one for ejection clearance of the molding, and the second opening for the removal of the runner and sprue system.

Parts which cannot be formed by the normal functioning of the press require separate or *loose details or cores* in the mold to hold inserts or to form threads or coring.

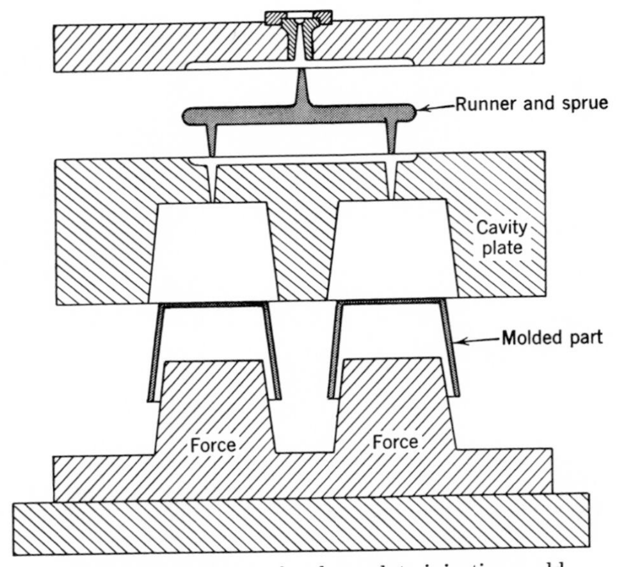

Fig. 17. A diagram of a three-plate injection mold.

These members are loose and are ejected with the molding, separated, and reinstalled in the mold. Duplicate sets are required for efficient operation.

Pneumatic or hydraulic cylinders may be mounted on the mold to actuate *horizontal coring* members. The air or hydraulic system is operated by the press or by an auxiliary air supply, and is controlled by switches and valving interlocked within the press movements for safety and sequential operation. By the addition of *angular core pins* engaged in sliding mold members it is possible to mold horizontal or angular coring without the necessity of costly loose details.

Automatic unscrewing for internal or external threads on parts molded at high production rates may be provided at relatively low cost by the use of a rack and gear mechanism, actuated by an hydraulic double-acting long-stroke cylinder. This operation must be sequentially timed within the press movement. Other methods of unscrewing threaded parts involve the use of an electric gear motor drive or friction mold wipers engaged on the periphery of the molding and actuated by double-acting cylinders after the mold has opened.

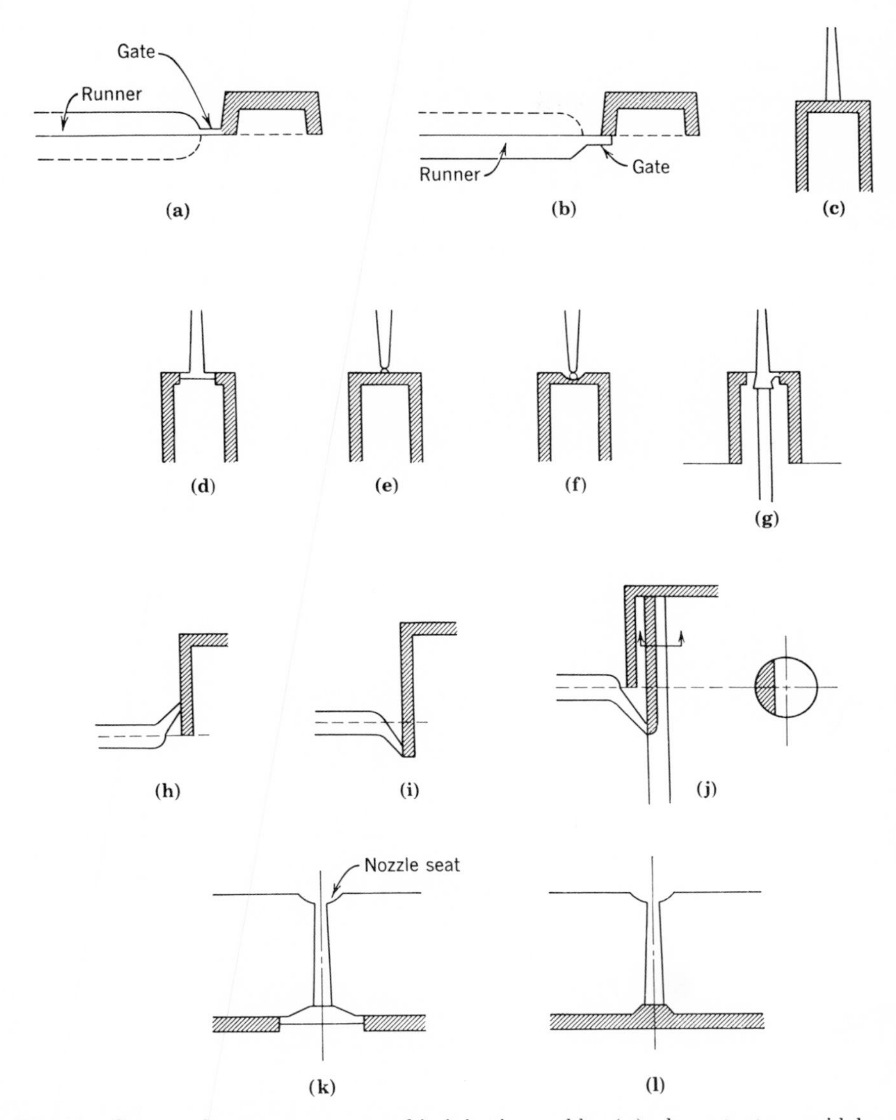

Fig. 18. Gates and runner systems used in injection molds: (**a**) edge gate, trapezoidal or round runner; (**b**) bottom gate, trapezoidal or round runner; (**c**) sprue gate; (**d**) disc gate; (**e**) pin gate; (**f**) depressed pin gate; (**g**) leg gate; (**h**) top cavity submarine gate; (**i**) bottom submarine gate; (**j**) submarine-plug gate; (**k**) sprue disc gate; (**l**) sprue gate.

Moldings with undercuts on the interior of the part can be made in a mold which permits *angular movement of the core.* Movement is actuated by the ejector bar and frees the metal core from the molding. Decorated parts on which the gate may not show are molded by incorporating the sprue and gate internally on the part, designing the *ejection system around the long sprue bushing,* and actuating the ejector movement by chain pulls or pull rods actuated upon press opening. Thin-walled and deep molded parts requiring uniform ejection are ejected by external bushings machined into a *stripper plate* and actuated by the ejector bar.

Runner Systems. The channels permitting the material to enter the gate areas of the cavities are called runners. Cross sections of runners are normally either full round or trapezoidal. Full round offers the best movement of material, but requires a duplicate machining operation in the mold, since both plates at the parting line must be cut. Trapezoidal runners are efficient, and are desirable in three-plate mold designs, in which sliding movements are required across the parting-line runner face. This design permits free movement of mold wipers for runner removal in automatic molding. Cavity layouts permitting balanced runner systems, ie, equal distance to all gating areas from the main sprue, are preferred for more uniform flow of material. Unbalanced systems for multicavity molds require reworking after pilot sampling to improve the flow and filling pattern.

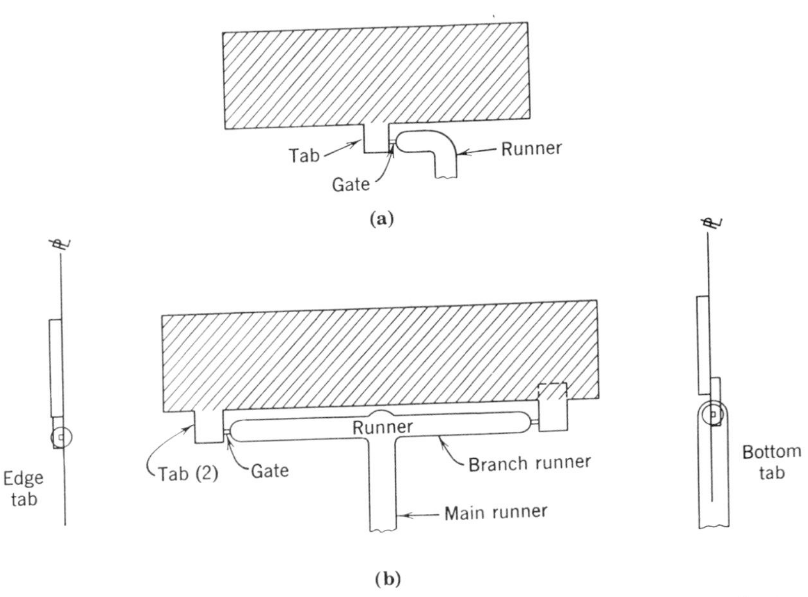

Fig. 19. (**a**) Single-tab gate (edge or bottom) and (**b**) double-tab gate (edge or bottom) used in injection molds. PL stands for parting line.

Gating. Various gating designs are used for injection molds, depending on product design and molding material used. Figures 18 and 19 show the many types of gates in use, as well as runner systems. In all cases, the dimensions of the gate land must be limited.

Cooling. The majority of injection-mold designs require channels to permit the circulation of cold water in order to extract heat from the mold and thus to harden the plastic. Other materials require the channels in order to obtain uniform mold temperatures. Mold frames are constructed with a series of channels connected so as to permit a uniform flow of cooling or heating medium, and thermostatically controlled to maintain a given temperature. Cooling may also be accomplished by direct channeling into the cavities and plungers for efficient operation close to the molding areas.

Proper cooling or coolant circulation is required for uniform repetitive mold cycling; auxiliary coolant circulation may be used in conjunction with the mold.

Mold Making

Molds, after engineering designing is completed, are manufactured by skilled craftsmen using the general types of metal-working equipment. Specifications for molds should include all details and their required tolerances. The use of commercially available mold bases, plates, frames, standard ejectors, and miscellaneous items is a good practice for manufacturing and future standardization of replacement items.

Advanced types of tools, such as the electrical discharge machine, and tape-controlled drill presses, milling machines, and jig borers, have made it possible to produce the many molds for the newer molding materials and sophisticated tooling designs that are necessary in the plastics industry.

A wide variety of mold surfaces are available which give dull, bright, glossy, satin, and various textured surfaces, as well as high polish and optical grades for specific applications. Hard chrome plating, highly polished, is generally recommended for

Fig. 20. Modern tool room. Courtesy Athenia Plastic Mold Corp.

Fig. 21. The processing of mold sections by duplicating, engraving, and milling machines. Courtesy Athenia Plastic Mold Corp.

Fig. 22. Contour-cutting lathe showing a formed plunger being machined. Courtesy Athenia Plastic Mold Corp.

Fig. 23. Grinding and preliminary gaging and inspection operations for mold processing. Courtesy Athenia Plastic Mold Corp.

thermosetting molds to ensure good molded finish, to enable easier ejection, and to minimize mold wear.

Figures 20, 21, 22, and 23 show a modern tool room specializing in plastic mold making. Proper inspection and quality control are necessary at the time of mold construction to ensure the quality of the final assembly.

Hobbing. This method of mold fabrication is commonly divided into two categories. In *cold hobbing*, a master hardened-steel hob, from tough steel such as listed in Table 1, is machined to the shape of the part, either as an exterior shape for the

Table 1. Steels for Injection, Compression, and Transfer Molds[a]

AISI type	Method of fabrication	Annealed hardness, Bhn[b]	Analysis								Comment	Normal work hardness, Rockwell C
			C	Mn	Ni	Cr	Mo	V	W	Si		
A-2	machine	212	1.00	0.60		5.25	1.10	0.25				59–62
A-4	machine	229	0.95	2.00		2.20	1.15			0.35	may be leaded	59–62
A-6	machine	235	0.70	2.00		1.00	1.00					58–60
S-1	machine	212	0.50			1.50			2.50			58–60
S-7	machine	197	0.50	0.70		3.25	1.50					55–57
H-11	machine	207	0.35			5.00	1.50	0.40				50–54
H-12	machine	217	0.35			5.00	1.50	0.40	1.50			50–54
H-13	machine	207	0.35			5.00	1.50	1.00				50–54
P-2	hob	103	0.07		0.50	2.00	0.20				carburize case	60–64
P-3	hob	116	0.10		1.25	0.60					carburize case	60–64
P-4	hob	121	0.07			5.00	0.75				carburize case	60–64
P-6	machine	207	0.10		3.50	1.50					carburize case	58–61
P-20	machine	300	0.35	0.80		1.70				0.50	prehardened	30–34
4150	machine	300	0.50	1.00		1.10	0.20				prehardened	30–34
420ss	machine	217	0.35	0.40		13.00					mold quality	50–52
						Master hob steel						
O-1	machine	202	0.90	1.00		0.50			0.50			59–61
A-2	machine	212	1.00	0.60		5.25	1.10	0.25				59–62
D-2	machine	212	1.50			12.00	1.00					58–61
S-1	machine	212	0.50			1.50			2.50		carburize case[c]	58–61
S-7	machine	197	0.50	0.70		3.25	1.50				carburize case[c]	60–62

[a] Courtesy Lindquist Steels Inc. (by T. McFadden, Jr.).
[b] Bhn is Brinell hardness number.
[c] Slightly carburized case may be used for added wear.

cavity, or, in some cases, as an interior shape for the plungers. The hob is placed in a press, and, using high pressures, is cold pressed into a premachined soft-steel blank contained in hardened-steel retainers. The hob is forced into the blank, displacing the steel to form the external shape of the hob.

This procedure is used to make multiple-cavity molds, as well as single-cavity molds, when the design dictates and economics favor machining of the master hob rather than machining of the cavity.

Suggested types of steel to be used for the master hob and hobbings are shown in Table 1.

Hot hobbing entails the pressure casting of mold sections in beryllium–copper alloy, and is the most reliable method of hobbing. The molten alloy is poured over the master hob, which is contained in a chamber to form the final casting. Pressure is applied by an hydraulic press to a force plug into the chamber, which forms the casting into a solid and dense state. The final casting is heat treated to produce good physical properties. Hot hobbing is most commonly used in thermoplastic molding.

Materials. The most widely accepted material for all types of molds is steel, with many types of alloys in use for particular applications. Molds for thermosetting polymers normally require steels that are hardened after fabrication. Molds for thermoplastic materials employ the same types of steel, with the addition of many prehardened classifications. Table 1 covers a summary of materials that are extensively used. Many other types are available and Table 1 is to be used as a reference only.

Other materials employed in the manufacture of molds for plastics are aluminum, brass, epoxy-modified castings, electroformed nickel, and beryllium–copper castings. Beryllium–copper alloy may be used for injection molds but does not have sufficient stability for other methods. Molds for thermosetting plastics must have the strength, hardness, and stability to withstand the pressures and the materials employed; injection molds can be made of prehardened, or cast mold materials, which are easier to machine.

Type of mold, sophistication of product design, material to be molded, type of processing, and production requirements are all factors in selecting the material. It is good judgment to process with the best materials available for a given designand employ materials of less high quality only as an expedient for prototypes or short runs. The overall costs in the manufacture of a mold are generally about 5% for engineering, approximately 15% for materials, and the remainder for processing.

Bibliography

Modern Plastics Encyclopedia, McGraw-Hill Book Co., Inc., New York.
Plastics Engineering Handbook, 3rd ed., Reinhold Publishing Corp., New York, 1960.
Plastics World, Cahners Publishing Co., Englewood, Colorado.
J. H. DuBois and F. W. John, *Plastics*, new and rev. ed., Reinhold Publishing Corp., New York, 1967.
J. H. DuBois and W. I. Pribble, *Plastics Mold Engineering*, rev. ed., Reinhold Publishing Corp., New York, 1965.

Sumner E. Tinkham
Tech-Art Plastics Company

RELEASE AGENTS

Release agents, also called abherents or parting agents, are solid or liquid films or granular solids that reduce or prevent adhesion between two surfaces. In the field of polymers, they are of importance in rubber and plastic processing, paper coating, and production of pressure-sensitive tapes. A number of factors influence adhesion of two materials to each other. The most important ones are penetration, chemical reaction and compatibility, surface tension, surface configuration, and polarity differences between the two materials. Two solid surfaces generally do not adhere to each other because wetting does not take place nor does penetration of one into the other. The only exception occurs when one of the surfaces is "tacky" or when chemical reaction takes place between the two surfaces at the interface. Frequently, high static charges can also lead to adhesion. Two smooth and glossy plastic surfaces adhere to each other also through the creation of a vacuum, which can be prevented by the use of release agents. Therefore, the use of release agents becomes of technical importance when a solid and a liquid, or, even more so, when a solid and a paste or dough form an interface and adhere to each other. For many centuries adhesion of a highly viscous material, such as paste or dough, was a problem in the home in baking and cooking. Abherents in the form of fats, oils, or solids like flour were used in order to prevent the sticking of dough to wooden kneading boards or to various metal baking dishes. With the greater industrial use of polymeric materials, both natural and synthetic, the commercial use of release agents has become widespread. Indeed, some industries which are of great importance today could not have developed without the availability of modern abherents. An example is pressure-sensitive tapes, which could not be unwound if the tape backing were not coated with a release agent.

Properties Required

Since many of the factors causing adhesion are of a chemical nature, one of the first requirements of a good release agent is complete chemical inertness toward the two materials whose adhesion is to be prevented (see also ADHESION AND BONDING). Adhesion is often due to opposing polarities of the surfaces, therefore the polarity relation of the release agent and of one or both of the surfaces in contact with it have to be taken into consideration in the choice of release agent. Besides these two factors, a physical property of great importance is good spreading ability, ie, low surface tension, so that the release agent will form a continuous film between the surfaces and in this way exclude any contact between the two materials. An extremely important factor in this action of release agents is temperature dependence. This dependence is so pronounced that a material which is an abherent at, for example, room temperature may act as an adhesive at elevated temperatures. A good example of this is polyethylene film, which is used as an abherent layer to prevent the adhesion of uncured rubber slabs or preforms to each other. This same polyethylene film, above its softening point at approximately 100°C, acts as an excellent hot-melt adhesive. Another factor to be considered is volatility. Water would fulfill many of the requirements of an abherent, but because of its vaporization and the increased water absorption of many materials at elevated temperatures, water in most cases is not suitable as a release agent. With knowledge of all these factors, it is possible to narrow considerably the choice of a release agent. It must (a) be chemically inert toward the two adhering materials, (b) have good spreading properties for one or both of the surfaces, (c) have low volatility at the temperature at which it is to be used.

In many cases adhesion is caused primarily by the physical configuration of the solid surface, namely, by its high porosity or roughness, which tends to anchor a viscous material into the surface. In this case a powdered solid material will serve well as a release agent by filling the pores or smoothing out the roughness of a surface. On the other hand, adhesion between two solid surfaces can also be caused by the smoothness (gloss) of the surfaces and thereby creation of a vacuum which holds the surfaces together. This being a mechanical adhesion, it can best be relieved by release agents that either roughen the surface of the polymer film (additives) or form a reasonably coarse interface that will allow the vacuum to be relieved.

It is obvious that a liquid, to be effective as an abherent between two materials, must spread so as to form a continuous film between the contacting surfaces. Therefore, an additional requirement for a good release agent is a low to medium viscosity at the temperature of application. There is an exception to this requirement when polymer melt flow is induced by high pressures. In this case a release agent of higher viscosity is more desirable so that it is not displaced by the polymer melt.

Methods of Application

As mentioned above one of the most important uses for abherents is at the interface of a solid and a dough or paste, eg, a polymer melt. Three basic methods for the application of a release agent in this case are: (a) spraying, brushing, or dusting a powdered solid or liquid abherent to the solid surface; (b) producing a permanent abherent by baking an abherent polymeric surface to the solid surface; or (c) incorporating the release agent in the polymer. In this latter case the abherent must be partly compatible with the polymer, at least at room temperature, and must exude to some extent at the melt temperature in order to be available at the interface where adhesion has to be prevented. Many polymers have a slight amount of tack even in the solid state. This property has caused considerable difficulties in the processing of some polymeric materials. A good example is given by polyethylene films, which tend to adhere to each other because of static electricity and also because of cold flow. Poly(vinyl chloride) films show the same behavior. Therefore, the incorporation of abherents into these polymers before they are processed into films is standard practice. In the case of poly(vinyl chloride) films, these additives are called antiblocking agents, whereas in the polyethylene field they have become known as slip agents, although in either case they really serve the same functions.

Industrial Fields Using Abherents

Rubber Processing. Release agents have attained considerable importance in the processing of rubber. Generally, both natural and synthetic elastomers have excellent adhesive qualities in the uncured state, yet they must be processed and shaped before they can be cured. There are a great variety of processing steps involved in the manufacture of rubber goods. Generally the procedure may be broken down as follows: First, elastomers are blended with various compounding ingredients, such as fillers, accelerators, vulcanizing agents, pigments, and other agents. This is usually accomplished in an internal mixer of the Banbury type. Then, this mixed batch of rubber is generally "sheeted" through a rubber mill to put it into the form of a sheet or slab, which may then be calendered or molded into thinner sheets, extruded into tubing, or formed into other preforms for molded rubber goods. During these processing steps that precede vulcanization, the rubber compounds would tend to stick to metal surfaces and even more so to each other. In order to prevent this sticking, coated papers and coated cloths have been used for many years to separate the individual layers of uncured rubber. Also, ever since rubber processing has become an industrial art, various release agents have been used to dust or spray onto the molds. Abherents have also been incorporated into the compounds to reduce tackiness. All of the various types of abherents have been and are still being used in the rubber industry. Of primary importance are the various metal stearates, stearic acid, oleic acid, microcrystalline and paraffin waxes and other synthetic waxes, such as the stearamides, *N*-substituted alkyl stearamides, and bis-stearamides (derived from diamines, eg, ethylenediamine), and ester waxes, such as montan wax. In rubber processing, as in many other fields in which release agents have been employed, the advent of the silicones has brought about major changes. Previously, molds for molding rubber goods or for curing rubber sheeting had been dusted after each cycle with zinc stearate, calcium stearate, and similar abherents. These have mostly been replaced by silicones in the form of water dispersions, which are sprayed onto the molds in dilutions of 1% or below. Whereas previous lubricants had led to buildup of crusts in the molds and had necessitated frequent cleaning, the silicones have eliminated this shortcoming completely. Another use of abherents in the rubber industry is the dusting of finished unfilled rubber goods which would be tacky without a surface application of abherent. In this case solid materials have been found to serve best; talcum, mica, finely dispersed silicates, or metallic stearates have proved advantageous (see also DRIERS AND METALLIC SOAPS).

Stearic acid, one to five parts per hundred of rubber, usually forms part of any rubber formulation. The function of this fatty acid is not completely clear. It is assumed that the stearic acid reacts with the zinc oxide present in every formulation during vulcanization; furthermore, that the zinc stearate thus formed, in turn, reacts with the accelerator, enabling the latter to promote the vulcanization. Without any doubt the addition of fatty acid also produces an abherent effect in this case (see also RUBBER COMPOUNDING AND PROCESSING).

Polymer Processing. The use of release agents is of the utmost importance in polymer processing. Some polymers have particularly great adhesive properties at or near their melting points and are, therefore, in greater need of abherents than others. Examples of these highly adhesive melts are polystyrene, some polyolefins, other hydrocarbon resins, poly(methyl methacrylate), and, to some extent, polyamides.

But even polymers with lower adhesive forces require the use of release agents in most cases. The polymers mentioned above, which are all thermoplastic in nature, are not anywhere near as adhesive as some thermosetting resins that have to be cured in contact with the mold surface. Examples of these latter are the polyesters, polyurethans, and epoxy resins.

The use of release agents in polymer processing includes all possible methods of application. Most polymeric compounds made by raw-material manufacturers already contain abherents. Furthermore, the processor who adds pigments and other ingredients to the raw resins also adds abherents at that stage. In many cases the fabricator who molds or extrudes thermoplastics will tumble some release agent or abherent onto the plastic powders or granules before melting them in the processing equipment. Even beyond this preparatory addition of release agents an application of abherents to the mold surfaces or metal surfaces of other processing equipment is frequently necessary.

A few specific examples may be cited to illustrate the application of abherents. In the injection molding of polystyrene, abherents are added when the raw material is processed. Pure polystyrene as it comes out of the polymerization process needs the addition of abherents to promote the conversion of the raw polymer into molding pellets or granules. This conversion, which is usually accomplished through an extruder, is greatly facilitated by the addition of small amounts of abherents, such as metallic stearates or stearic acid. The increase in throughput through the extruder, when materials containing abherents are compared to those not containing any, can be as much as 20%. These molding pellets or granules, after they have beeen cooled, are again tumbled with a release agent, which more or less adheres to the surfaces of the individual pellets. In this case a great variety of release agents can be applied. Zinc stearate, stearic acid, and stearamide are among the products most frequently used, but substances that are liquid at room temperature, such as butyl stearate, are also employed in many cases (see also MOLDING).

In the conventional injection-molding process unmelted plastic pellets are pushed by a hydraulic ram into a cylinder where they are gradually heated and melted. It is the function of the abherent to reduce the friction of the granules against each other and also against the metal surfaces of the cylinder in order to reduce the pressure needed to convey them forward into the melting zone. This polymer melt is then injected by the ram into a cold form or mold. Most injection-molding machines built during recent years use a plasticating screw either in conjunction with a ram or by accomplishing injection by reciprocating movement of the screw. This method has greatly reduced the need for release agents. Nevertheless, they can still reduce required pressure as well as the cooling cycle by faciiltating release of the molded item from the metal mold surface. In many cases additional release agents have to be applied to the mold surface in order to further facilitate the release of the molded plastic article from the metal mold. It can easily be seen that the function of release agents in facilitating the molding of a polymer will also reduce the temperature to which the material has to be heated for it to flow properly into the mold. Since high temperatures are detrimental to the stability of most polymers, the release agents therefore also have a protective function, and result in articles of greater strength and higher quality. It should be noted that silicones are not used as abherent additives to polymers. The reason is that silicones are so powerful as abherents and so inert and noncompatible that their addition to polymers would counteract the cohesive forces and destroy the homogeneity of the material.

Silicones make very effective release agents when applied to the mold in the injection or compression molding of plastic materials, but they must be used with great caution, because molded polymeric items often have to be adhered to other surfaces, or lacquers have to be applied to them for decorative purposes. Silicones remaining on the surface of these molded pieces would seriously interfere with these aftertreatments.

In the processing of rigid and flexible poly(vinyl chloride), release agents are also of great importance, although poly(vinyl chloride) does not have as strongly adhesive a character in the melted state as polystyrene. In the manufacture of films from poly(vinyl chloride), the polymer compounds are generally blended in a cold or a slightly heated blender. They are then melted and fused on a heated rubber mill or in an internal mixer. This melt is then fed into a calender consisting of three or four heated rolls through which the poly(vinyl chloride) compound is squeezed down to the thickness desired in the finished film. For calendering, abherents must be carefully compounded with the polymer, so that the melt will follow the proper rolls and release from the ones that should not be followed. Frequently the use of one abherent is not sufficient, and a combination of two or more must be applied. An additional function of the abherent lies in its action as an antiblocking agent. Whereas small amounts of stearic acid provide excellent release from metal surfaces, other abherents, such as calcium stearate, lead stearate, and stearyl amides, act as a combination of release and antiblocking agents. These latter abherents also greatly influence the surface appearance of the calendered film, providing such desirable properties as gloss and smoothness.

In the extrusion of poly(vinyl chloride), a process that consists in melting a granulated or powdered poly(vinyl chloride) compound between a cooled rotating screw and a heated cylinder and in then pushing this homogenized heated melt through a shaping orifice, the addition of release agents helps to release the compound from the rotating screw and in this fashion greatly increases the speed with which the compound is conveyed through the extruder cylinder. The release agents also reduce the possibility of thermal breakdown, which is critical with poly(vinyl chloride). Since the shaping orifice or die is usually the hottest part of the extruder, easier release from this hot and normally glossy surface is provided by coating the die surface with a silicone paste or grease prior to extrusion. This serves to fill in the pores of the metal and to prevent carbonization of the polymer melt as a result of prolonged contact with the hot metal. Coating of the die surface with release agent, however, is no substitute for the addition of various abherents to the compound itself.

In the casting of thermosetting plastic materials, such as polyesters and epoxy resins, it is most essential to provide easy release from either the male mold over which such a compound is cast or the female mold and flexible bag in which it is molded. Matched metal molds are also sometimes used to shape these compounds. Epoxy resins (qv) are known to be outstanding adhesives. Where release from a mold surface must be provided, the use of release agents in one fashion or another is an absolute necessity. It is interesting to note that, in the field of casting thermosetting materials, the use of polymeric films as abherents, in the form of extruded or cast films, has been found advantageous. Polyamide or poly(vinyl alcohol) films have proved highly successful in this application. In addition to these films, sprayed-on silicones or waxes are also frequently employed.

Paper Coating and Pressure-Sensitive Tapes. Adhesive-coated papers, cloth, and plastic films have attained great industrial importance during the last twenty years. All these materials, after being coated with adhesive and dried, must be wound up in rolls. Release agents are needed to prevent the rolled-up adhesive coating from sticking to the backing material. These applications make very severe demands on abherents, and considerable work has been done in this field in order to develop the proper materials.

Here again, as in so many applications, silicones are outstanding. Silicone-coated papers have attained great commercial importance and are being produced by a large number of specialty paper manufacturers. These papers, which are coated with the silicone abherents on either one or both sides, are used as interleafing papers for various sticky substances like uncured rubber, whereas the papers coated on one side only are used primarily as the base material for pressure-sensitive and other adhesive paper tapes. Before the introduction and acceptance of silicone coatings, wax–polyethylene coatings, simple wax coatings, and zinc stearate or talcum dusting had been used to a great extent.

In an application combining both paper coating and polymer processing, poly(vinyl chloride) plastisols, which are dispersions of finely powdered polymer in liquid plasticizers, are frequently cast into sheets or films that are then fused into a polymeric compound. Originally, these films were cast onto stainless-steel continuous belts, an operation that entailed a considerable initial investment as well as replacement cost. These films can now be cast and fused on abherent-treated papers, and the films are peeled off the papers after fusion.

Abherent-treated papers are also used widely in the general packaging of sticky materials, such as chemicals, foods, etc. Practically all classes of abherents are used in the packaging field.

Classes of Release Agents

Waxes (qv). Both natural and manufactured waxes are used as release agents. Natural waxes that have gained importance in this field include paraffin and microcrystalline waxes (both types are petroleum products); waxes of vegetable origin, such as carnauba or candelilla wax; and waxes of animal origin, such as spermaceti.

The petroleum-based waxes have the disadvantage of relatively low melting points, which may produce stickiness rather than release near the melting point. These waxes also are subject to oxidation at high temperatures. The vegetable waxes, like carnauba wax, are excellent release agents but, being natural products, have the disadvantage of variation in color and price. High price also has been limiting the use of spermaceti wax.

Synthetic or manufactured waxes have attained greater importance for use as release agents. Practially all aliphatic alcohols from the C_{10} up have found use as release agents. The same can be said about fatty acids above C_{12}. The fatty acid that has attained the widest use as a release agent is stearic acid, which is available highly purified at a reasonable price. It has a sharply defined melting point and good wetting properties. Since stearic acid also has a very limited compatibility with organic polymers, it has been found an efficient abherent in a great number of applications. Examples are the internal lubrication produced (by addition to the formulation) in poly(vinyl chloride) and in styrene polymers and copolymers (see also LUBRICANTS). The use of stearic acid in rubber compounds, where it not only serves as an abherent but performs other functions as well, has been covered. See also ACIDS AND DERIVATIVES, ALIPHATIC.

The glyceryl stearates and various glycol stearates comprise another group of synthetic waxes that are employed as release agents in a multitude of applications. Two important members of this family are (a) glyceryl monostearate (α-stearin), used as an additive in poly(vinyl chloride) compounds for the manufacture of sheeting and film and as an additive in polybutylenes (Vistanex, Enjay Chemical Co.), which otherwise have exceptional adhesive properties, and (b) diethylene glycol monostearate, used as an incorporated abherent in rubber processing and in poly(vinyl chloride) compounds and as an abherent coating on paper. Another interesting group in this family of glyceryl fatty acid esters are the hydrogenated oils. Of particular industrial interest is fully hydrogenated castor oil, which is sold under the trade name of Opalwax (Baker Castor Oil Co.). Chemically, Opalwax is glyceryl tri(hydroxystearate). This wax has an exceptionally high melting point and is used as an abherent in rubber compounds, coated fabrics, and papers. Probably the most important single release agent in the family of synthetic waxes, known under various trade names (Acrawax C, manufactured by Glyco Chemicals, Inc.; Advawax 280, manufactured by Advance, division of Carlisle Chemical, Inc.), is reported to consist primarily of ethylene bis-stearamide. This is an amorphous wax of good color, excellent heat stability, and a melting point of 140–143°C, probably the highest of all the waxes. This wax, having such a high melting point, can be obtained and shipped in a very finely pulverized form, which makes it more convenient to apply than most waxes. It also has extremely limited solubility and compatibility and, therefore, can be used in a great number of systems. Acrawax C has been found useful as a release agent in almost all types of polymer processing and, in many cases, acts in the finished polymer film or sheet as an antiblocking agent, or slip agent, preventing adhesion of two layers of film to each other.

Metal Salts of Fatty Acids. The metal salts of fatty acids and primarily of stearic acid (see DRIERS AND METALLIC SOAPS) have acquired an important industrial position as release agents. They are high-melting solids, usually available in a powdered form; these metal salts are applied primarily as dusts, but also in many cases are incorporated into polymeric compounds as internal lubricants or release agents. In rare instances water dispersions of these metallic salts have also been applied. Zinc stearate, for example, is available in a water dispersion and has been used in this form in rubber processing. The most important metallic salts used as release agents are calcium stearate, zinc stearate, lead stearate, magnesium stearate, and aluminum stearate. Which metallic salt is chosen for a specific application depends primarily on the polymers and other surfaces involved. Calcium and lead stearate are the dominant release agents in poly(vinyl chloride) processing, where both also have heat stabilizing effects. Zinc stearate is substituted for lead stearate in specific cases where nontoxicity is a requirement, but it does not have anywhere near the stabilizing effect of lead stearate. Calcium stearate is probably the most effective abherent in poly(vinyl chloride).

As an abherent in polystyrene, zinc stearate has found the greatest use. Aluminum stearate is also effective with many polymers. In the field of rubber processing, aluminum and magnesium salts are the preferred release agents. The reasons for the superior functioning of one metallic salt over another in various applications have been determined empirically. Frequently the choice of the right stearate can be based on melting point, insolubility, and particle size to which the stearate can be ground or precipitated.

Metallic salts are used indirectly as release agents where a low-melting wax is the primary abherent and the metallic salt is employed to increase the melting point of the mixture. Thus, blends of aluminum stearate and paraffin wax and also of zinc stearate and stearic acid have been made in order to adjust melting points and physical properties of the primary abherent.

Polymeric Release Agents. In this section various types of polymers used as release agents are discussed.

Poly(vinyl Alcohol) (see VINYL ALCOHOL POLYMERS). This polymer, which is completely water-soluble and incompatible with practially all organic polymers, has a variety of uses as a release agent. It is applied as a coating from a water solution, or in the form of a cast or extruded film. One of the major applications is in the molding or lay-up forming of polyesters and epoxy resins.

Polyamides (qv). Since polyamides are insoluble in most of the commonly used solvents, polyamide films are extruded. These are used with polyester and epoxy resin lay-ups and are draped over the metal or plaster mold in order to effect release of the cured formed shape from the pattern.

Polyethylene (see ETHYLENE POLYMERS). This polymer is used as an abherent film (in the form of extruded tubular or flat film) in the processing and shipping of uncured rubber, and as a paper laminate in the packaging of sticky materials.

Silicones (qv). Silicones represent the most important class of release agents despite their rather recent origin. The commercially useful silicone release agents are all polymeric in order that they may have high boiling points and therefore low volatilities at room temperature, heat resistance, and resistance to oxidation. The higher the molecular weight, the greater the chemical resistance and, of course, also the viscosity.

Prices of the silicones are in the range of about ten times the cost of other release agents, yet in actual use they are in many cases considerably cheaper, because of the very small amounts needed (often they are used in highly diluted solutions or water dispersions below 1% silicone content), and also because they do not build up on metal surfaces as metal stearates do. As a result, they save considerable time in the cleaning of molds. In the field of polymer processing the only word of warning that applies to the use of silicones is that their abherent properties are too good. In many cases, where surface treatments in the form of printing inks or adhesives must be applied later, the silicones will interfere with adhesion.

The silicones are used in three forms: fluids, resins, and greases.

Silicone Fluids. Commercially, the most important group of fluid abherents are based on dimethyl silicone. They can be applied full strength, in solution, or in a water emulsion. These silicone fluids have almost every property required of a release agent: (a) excellent heat resistance and stability; (b) low surface tension; (c) great chemical inertness, especially toward organic materials and polymers; (d) colorless appearance and nonstaining qualities; and (e) physiological inertness. Their outstanding properties, together with the ease with which they can be applied, make the silicones ideal release agents in many instances. In applications such as metal processing and glass molding, where extremely high temperatures are involved, the silicones actually are the only release agents that will allow continuous use with great ease.

Silicone Resins. These are cured by heat and catalysts, which usually are metallic salts or organic amines. Crosslinking, of course, is favored by a higher functionality of the repeating unit. Whereas the silicone fluids usually possess two carbon-containing groups to each silicon atom in the chain, the repeating units for resins usually contain only one carbon group on each silicon atom, the rest being hydroxyl groups.

Silicone resins are manufactured and sold in solution in organic solvents, since solution increases their shelf life. They can be applied by spraying, dipping, or brushing. They are used as release agents primarily on metal surfaces and have found their most important application in baking pans in commercial bakeries. Here again the importance of the silicone resin coating lies in the extreme heat resistance of the fully cured films, which can be used up to approximately 400°F. A paint formulated from silicone resin with powdered aluminum pigment can attain heat resistance up to 500°F.

Silicone Greases. These are silicone fluids thickened with either lithium octoate or with a silica gel, such as Cab-O-Sil. The greases have the advantage of not running off too easily, even at elevated temperatures, and therefore acquiring somewhat more permanence than a silicone fluid would when applied to metal or other surfaces.

Fluorocarbon Polymers (see also FLUORINE-CONTAINING POLYMERS; TETRAFLUOROETHYLENE POLYMERS). The fluorocarbon polymers are available in water dispersions, but their general use as release agents has been retarded by their very high cost.

They do have the properties required of good release agents, namely very high chemical inertness and high heat resistance. They are employed as abherents in the form of films and sheets, primarily where permanent gasketing between two sticky surfaces is required. Their use, of course, is very limited since a fluorocarbon film or sheet sells for well above $10.00 per pound. One application of polytetrafluoroethylene, as mentioned previously, is as a coating for metal frying pans for home use. This coating is applied as a dispersion, which is dried and fused at approximately 250°C.

Inorganic Compounds. Inorganic compounds constitute the oldest release agents known. Owing to their insolubility, they are used strictly in the form of powders, which have abherent properties, generally because of their flakelike crystal structure. The most important representative of this class of release agents are talcum and mica. These are always applied as a fine powder sprayed or dusted onto a surface to prevent adhesion between that surface and metal or similar surfaces. They are frequently blended with metal stearates to improve their abherent action. A major application for these inorganic abherents is the dusting of soft rubber products, eg, tubing, rubber bands, and soft rubber sheets. Another example of the use of talcum as an abherent is the dusting of freshly printed paper or polymeric film so that it may be wound into rolls before the inks are completely dried, without transferring the inks to the backs of the web in the roll.

Bibliography

H. Bennett, ed., *Commercial Waxes*, Chemical Publishing Co., Inc., New York, 1956.

Dow Corning Corp. Bulletins 5-116, 5-115b, 5-111, U-5-100, 8-605.

J. W. Keil, "Silicone Paper Coatings," *Tappi* **41** (6) (June 1958).

J. W. Keil, D. L. Leedy, and L. H. Reinke, "Silicon Release Coatings for Paper," *Paper, Film Foil Converter* (August 1958).

R. N. Meals and F. M. Lewis, *Silicones*, Reinhold Publishing Corp., New York, 1959.

A. R. Morse, "How to Select the Proper Mold Release," *Plastics Technol.* **13**, 43 (April 1967).

George P. Kovach
Koro Corporation

INDEX

Abherents, 207
Assembly, 159
Autoclaves, 172, 178
Auxiliary procedures, 145

Bag molding, assembly lines in, 173
  cure cycles in, 179
  description of, 167
  design for, 181
  equipment, 171
  finishing in, 179
  materials for, 169
  processes, 175
  resins for, 169
  tooling for, 173
Biaxial rotation, 129
Blow molding, cold tube, 112
  description, 89
  extrusion in, 97, 116
  injection in, 112
  molds for, 91, 99
  neck in, 107
  parison, 101, 107, 119
  problems in, 117
  rotary wheel, 109
  swell in, 120
  temperature control in, 95, 119
  testing of containers, 122
Breathe and dwell, 20

Cavity design, 32
Cold molding, 16, 41
Compression molding, accessories, 26
  applications, 44
  description, 11
  economics, 46, 48
  effect of, pressure, 21
    temperature, 19
    time, 20
  of elastomers, 16
  history, 11
  machines for, 17, 23
  mold design, 17
  molding compounds for, 18
  part design, 17, 22
  presses, 23
  process, 12, 17
  speeds, 22
  of thermoplastics, 16
  of thermosetting resins, 188
Containers, 122
Cyclone separator, 148

Decorating, 159
Definitions, 1
Deflashing, 28, 156
Diaphragm, 167
Die setting, 153
Drying, 152

Ejection, 27, 184
Encapsulation, 45
Extrusion, in blow molding, 97, 102
  continuous, 105
  intermittent, 102
  manifold, 102

Finishing, 156
Flash mold, 13
Flash removal, 28, 156
Fluorocarbon polymers, as release agents, 215

Gating, 184, 202
Glossary, 1

Hobbing, 204

Injection blow molding, 112, 114
Injection molding, auxiliary procedures in, 54, 82
  color dispersion in, 73
  cycles, 51, 65
  description, 51
  design of screw for, 62
  of elastomers, 57
  machinery, 52, 59
  nozzles, 71, 74
  offset injection, 59
  rotary, 63
  straight injection, 58
  of thermoplastics, 53
  of thermosetting materials, 51, 72
Injection transfer, 40
Insert molding, 81

Jet molding, 57

Lester cylinder, 55

Mandrel head, 99
Metal salts, as release agents, 213
Molded neck, 107
Mold-handling systems, 133
Molding machines, automatic, 36, 38
Mold release, 133, 170, 207
Molds, automatic, 198
  for blow molding, 91
  break-in procedures for, 35
  for compression molding, 192
  construction, 203
  cooling of, 202
  design of, 184, 195, 199
  ejection systems in, 81
  gating in, 79
  heating of, 137
  for injection molding, 75
  lubrication of, 154
  maintenance of, 35, 187
  materials for, 92, 206
  for melamine dinnerware, 45
  plungers in, 197
  reciprocating, 104
    screw, 197
  for rotational molding, 127
  runner systems in, 76, 202
  screw closures in, 45
  sprueless, 80
  sprues in, 76
  temperature control for, 153
  for thermoplastics, 197
  for thermosets, 188
  for transfer molding, 194
  types of, 183
  venting of, 132
Multicolor molding, 82

Nomenclature, 1

Parison, 101, 107, 119
Parting agents, 207
Plastication, 53, 54, 67, 69
Plunger machines, 52, 60, 69
Plunger molding, 13, 14, 43
Pneumatic loader, 146
Polyamides, as release agents, 214
Polyolefins, molding of, 126
Poly(vinyl alcohol), as release agent, 214
Postcuring ovens, 29
Postmolding processes, 145
Preheaters, 28
Preplastication, 53, 69
Pressure, 21
Purging, 155

Ram injection, 42
Raw materials, drying, 152
  handling, 145
  loading, 145
Reclamation, 160
Recycling, 36
Reinforced plastics, 167
Reinforcements, 169
Release agents, application methods, 208
  classes of, 212
  description of, 207
  inorganic, 215
  in paper coating, 211
  polymeric, 214
  in polymer processing, 209
  in rubber processing, 209
  in tapes, 211
Rotational molding, description of, 125, 127
  drive systems for, 135
  economic aspects of, 141
  heating in, 137
  internal mold blanketing in, 139
  machines for, 133
  molds for, 127
  resins for, 126

Safety, 37
Scrap, 160
Screw plunger, 40
Screw-plunger preplastication, 53
Screw-type machines, 59
Silicones, as release agents, 214
Steel for molds, 205

Testing, 122
Thermoplastics, in compression molding, 16
  in injection molding, 57
Thermosetting injection, 38
Thermosetting resins, molding of, 19, 167
  molds for, 188
Tooling, 30
Transfer molding, 11, 13, 14, 38, 188
Transfer-molding presses, 26

Vacuum hopper, 147
Vacuum pump feeding, 149
Vacuum venting, 29
Venting, 29, 184
Vertical hydraulic ram, 25
Vinyl dispersions, 126

Waxes, as release agents, 212